Memoirs of the American Mathematical Society
Number 354

Thomas L. Miller, Robert F. Olin,
and James E. Thomson

Subnormal operators and representations of algebras of bounded analytic functions and other uniform algebras

Published by the
AMERICAN MATHEMATICAL SOCIETY
Providence, Rhode Island, USA

September 1986 · Volume 63 · Number 354 (second of 3 numbers)

MEMOIRS of the American Mathematical Society

SUBMISSION. This journal is designed particularly for long research papers (and groups of cognate papers) in pure and applied mathematics. The papers, in general, are longer than those in the TRANSACTIONS of the American Mathematical Society, with which it shares an editorial committee. Mathematical papers intended for publication in the Memoirs should be addressed to one of the editors:

Ordinary differential equations, partial differential equations, and applied mathematics to JOEL A. SMOLLER, Department of Mathematics, University of Michigan, Ann Arbor, MI 48109

Complex and harmonic analysis to LINDA PREISS ROTHSCHILD, Department of Mathematics, University of California at San Diego, La Jolla, CA 92093

Abstract analysis to VAUGHAN F. R. JONES, Department of Mathematics, University of California, Berkeley, CA 94720

Classical analysis to PETER W. JONES, Department of Mathematics, Box 2155 Yale Station, Yale University, New Haven, CT 06520

Algebra, algebraic geometry, and number theory to LANCE W. SMALL, Department of Mathematics, University of California at San Diego, La Jolla, CA 92093

Geometric topology and general topology to ROBERT D. EDWARDS, Department of Mathematics, University of California, Los Angeles, CA 90024

Algebraic topology and differential topology to RALPH COHEN, Department of Mathematics, Stanford University, Stanford, CA 94305

Global analysis and differential geometry to TILLA KLOTZ MILNOR, Department of Mathematics, Hill Center, Rutgers University, New Brunswick, NJ 08903

Probability and statistics to RONALD K. GETOOR, Department of Mathematics, University of California at San Diego, La Jolla, CA 92093

Combinatorics and number theory to RONALD L. GRAHAM, Mathematical Sciences Research Center, AT&T Bell Laboratories, 600 Mountain Avenue, Murray Hill, NJ 07974

Logic, set theory, and general topology to KENNETH KUNEN, Department of Mathematics, University of Wisconsin, Madison, WI 53706

All other communications to the editors should be addressed to the Managing Editor, WILLIAM B. JOHNSON, Department of Mathematics, Texas A&M University, College Station, TX 77843-3368

PREPARATION OF COPY. Memoirs are printed by photo-offset from camera-ready copy prepared by the authors. Prospective authors are encouraged to request a booklet giving detailed instructions regarding reproduction copy. Write to Editorial Office, American Mathematical Society, Box 6248, Providence, RI 02940. For general instructions, see last page of Memoir.

SUBSCRIPTION INFORMATION. The 1986 subscription begins with Number 339 and consists of six mailings, each containing one or more numbers. Subscription prices for 1986 are $214 list, $171 institutional member. A late charge of 10% of the subscription price will be imposed on orders received from nonmembers after January 1 of the subscription year. Subscribers outside the United States and India must pay a postage surcharge of $18; subscribers in India must pay a postage surcharge of $15. Each number may be ordered separately; *please specify number* when ordering an individual number. For prices and titles of recently released numbers, see the New Publications sections of the NOTICES of the American Mathematical Society.

BACK NUMBER INFORMATION. For back issues see the AMS Catalogue of Publications.

Subscriptions and orders for publications of the American Mathematical Society should be addressed to American Mathematical Society, Box 1571, Annex Station, Providence, RI 02901-1571. *All orders must be accompanied by payment.* Other correspondence should be addressed to Box 6248, Providence, RI 02940.

MEMOIRS of the American Mathematical Society (ISSN 0065-9266) is published bimonthly (each volume consisting usually of more than one number) by the American Mathematical Society at 201 Charles Street, Providence, Rhode Island 02904. Second Class postage paid at Providence, Rhode Island 02940. Postmaster: Send address changes to Memoirs of the American Mathematical Society, American Mathematical Society, Box 6248, Providence, RI 02940.

Printed in the United States of America.

The paper used in this journal is acid-free and falls within the guidelines established to ensure permanence and durability.

TABLE OF CONTENTS

Library of Congress Cataloging-in-Publication Data

Miller, Thomas L. 1952–

Subnormal operators and representations of algebras of bounded analytic functions and other uniform algebras.

(Memoirs of the American Mathematical Society, ISSN 0065-9266; no. 354)

"September 1986, volume 63, number 354 (second of 3 numbers)."

Bibliography: p.

1. Subnormal operators. 2. Banach algebras. 3. Representations of algebras.
I. Olin, Robert F., 1948– . II. Thomson, James E., 1948– . III. Title. IV. Series.

QA3.A57 no. 354 [QA329.2] 515.7'246 86-17381

ISBN 0-8218-2415-5

ABSTRACT

Let $H^\infty(G)$ denote the algebra of bounded analytic functions on a bounded region G. Let $\pi: H^\infty(G) \to \mathcal{B}(\mathcal{H})$ be a continuous algebra homomorphism with $\pi(1)=1$ and such that $S \equiv \pi(z)$ is a subnormal operator. If S is pure, then π is weak-star, weak-star continuous and unique. Under various hypotheses we describe the spectrum and essential spectrum of $\pi(f)$ for f in $H^\infty(G)$.

Let G be the open unit disc and let S be multiplication by z on $L^2(m)$ where m is Lebesgue measure on the unit circle. In the setting of the paragraph above we may assume that the range of π is contained in $L^\infty(m)$. Our basic structure theorem establishes a one-to-one correspondence between such π's and certain measures on the maximal ideal space of H^∞. The mapping π can be onto. It is an isometry if and only if it is one-to-one and has closed range if and only if $\pi(f)=f$ for each f in H^∞. The mapping π is one-to-one if and only if there exists a measurable set E with positive measure such that $(\pi f)|_E = f|_E$ for each f in H^∞. Theorems for more general regions are also obtained.

In the last chapter these results are generalized to the following setting: let Y and Z be compact spaces. Let μ be a probability measure on Z and p a continuous map of Y onto Z. We obtain a structure theorem similar to the one mentioned above for those representations π of C(Y) into $L^\infty(\mu)$ such that $\pi(h \circ p)=h$ for all $h \in C(Z)$.

AMS (MOS) subject classifications (1980). Primary 47B20, 46J15; Secondary 47A67, 30E25, 47C99.

Key words and phrases. Subnormal operator, representation, weak-star topology, maximal ideal space, H^∞, Banach algebra, Gelfand transform.

Subnormal Operators and Representations of Algebras of Bounded Analytic Functions and other Uniform Algebras

Thomas L. Miller[1],

Robert F. Olin[2]

and

James E. Thomson[2]

CHAPTER I

INTRODUCTION

Let G be a bounded domain in the plane $\mathbb{C}$ and let $H^\infty(G)$ denote the Banach algebra of bounded analytic functions on G. Let χ denote the function whose value at λ is λ for every $\lambda \in \mathbb{C}$. This paper is concerned with the theory of the continuous algebra homomorphisms from $H^\infty(G)$ into $\mathcal{B}(\mathcal{H})$ that send 1 to 1 and χ to S where S is a subnormal operator acting on a separable Hilbert space $\mathcal{H}$. The Banach algebra $\mathcal{B}(\mathcal{H})$ consists of the bounded operators on $\mathcal{H}$.

Received by the editors February 28, 1986.

[1]Some of the results in Sections 2 and 3 appear in the first author's Ph.D thesis written under the supervision of Robert Olin.

[2]The last two authors were partially supported by a grant from the National Science Foundation during the preparation of this paper.

Even the reader whose interest does not reside in the structure of subnormal operators may still find some interesting results in this work (in particular, Chapters 5, 6 and 7). For example, suppose μ is a probability measure whose support, denoted spt μ, is contained in ∂G, the boundary of G. Let S be the normal operator on $L^2(\mu)$ given by multiplication by χ; i.e., $S=M_\chi$ where

$$M_\chi f = \chi f$$

for all $f \in L^2(\mu)$. If π is a continuous algebra homomorphism of $H^\infty(G)$ into $\mathcal{B}(L^2(\mu))$ with $\pi(1)=1$ and $\pi(\chi)=M_\chi$, then the range of π, ran π, is contained in the commutant of M_χ, denoted $\{M_\chi\}'$. Since this last algebra equals $\{M_g\colon g \in L^\infty(\mu)\}$ and, for each $g \in L^\infty(\mu)$, one has $\|M_g\| = \|g\|$, we may view π as a representation into $L^\infty(\mu)$. (For $g \in L^\infty(\mu)$ the definition of the operator M_g on $L^2(\mu)$ is the obvious one; i.e.,

$$M_g f \equiv gf$$

for all $f \in L^2(\mu)$.) Question: Given any such measure μ, are there any representations $\pi\colon\ H^\infty(G) \to L^\infty(\mu)$ that send χ to χ? The answer is yes. There are many. What are they? That is, describe how they arise. We will give a classification theorem that answers this last question (consult Theorem 59).

In the last chapter we show how the problems mentioned in the last paragraph are special cases of the following problem. Let Y and Z be compact spaces. Suppose μ is a probability measure on Z and p is a continuous map of Y onto Z. Describe those representations π of C(Y) into $L^\infty(\mu)$ such $\pi(h \circ p) = h$ for all $h \in C(Z)$. Our answer to this problem describes a very natural one-to-one correspondence between the extreme points of the set of measures ν on Y such that $p(\nu)=\mu$ and the set of representations π mentioned. (We would like to thank C. Foias for suggesting that our methods in Chapter Five might be general enough to carry out this latter description. Those readers who like abstraction first and examples second should read Chapter Eight before they read Chapter

Five.)

From now on, we shall refer to a continuous algebra homomorphism π from $H^\infty(G)$ into $\mathcal{B}(\mathcal{H})$ that sends 1 to 1 and χ to S (where S is a subnormal operator) as a unital representation.

Some remarks seem in order as to why the theory of unital representations has some importance to that of subnormal operators. One of the tools that has been used successfully to discover the structure of the lattice of invariant subspaces of a subnormal operator S is the algebra homomorphism from $P^\infty(\mu)$ to $\mathcal{A}(S)$, described in [11]. If S acts on the Hilbert space $\mathcal{H}$ and its minimal normal extension N is defined on $\mathcal{K}$, then the scalar spectral measure for N is denoted by μ. (An excellent account for the general theory of subnormal operators is contained in [10].) Define $P^\infty(\mu)$ as the weak-star closure of the polynomials in $L^\infty(\mu)$ (the dual of $L^1(\mu)$), and $\mathcal{A}(S)$ as the weak-star closure of the polynomials (in the variable S) in $\mathcal{B}(\mathcal{H})$. Recall that $\mathcal{B}(\mathcal{H})$ is the dual of the trace class operators.) It turns out that $\mathcal{A}(S)$ is weakly closed and the weak operator topology and the weak-star topology agree on this algebra [33].

If $P^\infty(\mu)$ is antisymmetric; i.e., every real-valued function in it is constant, then $P^\infty(\mu)$ is isometrically isomorphic and weak-star homeomorphic to the algebra $H^\infty(G)$ for a suitably chosen domain G. (The regions G that arise in this fashion are characterized in [32].) The space $H^\infty(G)$, for any region G, is the dual of a separable Banach space [36]; a sequence $\{f_n\}$ in $H^\infty(G)$ converges weak-star if and only if it is bounded and converges pointwise everywhere on G. Therefore, the algebra homomorphism referred to earlier is an example of a unital representation that is an isometry and a weak-star homeomorphism.

Looking at the other techniques used in [5,33] to study subnormal operators (or those techniques used in [2,6,8] to study other classes of operators), one realizes the existence of a unital representation π in the $P^\infty(\mu)$ case) implies some important information about the lattice of invariant subspaces for S. (For a specific result, consult Theorem 3.2 in [8].)

One is naturally led into a vast array of problems; this paper should be viewed as an initial assault on this mound. We have taken the liberty of asking many questions which we were unable to resolve. They are scattered throughout this work. There are many questions left that we did not ask that are begging to be answered.

Our probing has been directed in two ways. On one hand, we have let the problems of existence and uniqueness dictate the course of investigation. On the other hand, we have let the problems related to the functional calculus and continuity steer our inquiries. An example of our journey along the first course has already been sketched in the second paragraph. If π is a unital representation of $H^\infty(G)$ into $\mathcal{B}(\mathcal{H})$, then in a natural way, π describes a functional calculus. Our objective in this light (the second course) has been twofold; describe the spectral mapping theorems associated with π, and discuss the weak-star continuity of π (does it hold?).

Clearly the latter investigation is motivated from the results in [11], as we mentioned earlier. There is another motivating source for this inquiry.

Example 1. Let G be a bounded domain in $\mathbb{C}$ and let μ be planar Lebesgue measure restricted to G. Let $\mathcal{H}$ be the space of analytic functions on G that belong to $L^2(\mu)$. It follows from [27] that $\mathcal{H}$ is a closed subspace of $L^2(\mu)$ and that M_χ is a bounded subnormal operator on $\mathcal{H}$. Furthermore, $\{S\}'$, where $S=M_\chi|_{\mathcal{H}}$, consists of those multiplication operators M_ψ, where $\psi \in \mathcal{H} \cap L^\infty(\mu) = H^\infty(G)$. Clearly then, the map defined by

$$\pi(f) = M_f \quad \text{on } \mathcal{H}$$

is a unital representation. In [3,4] the spectral mapping theorems associated with this particular unital representation are the focal points.

Recall that a point $\lambda \in \partial G$ is inessential if there is a $\delta > 0$ such that every $f \in H^\infty(G)$ extends analytically to the open disc $\Delta(\lambda,\delta)$. The remaining points on ∂G are called essential boundary points. If π is a unital representation defined on $H^\infty(G)$ and λ is an inessential boundary point, then π can be extended to $H^\infty(G \cup \Delta(\lambda,\delta))$ via the formula

$$\tilde{\pi}(h) \equiv \pi(h\,|_G)$$

for all $h \in H^\infty(G \cup \Delta)$. Thus, there is no loss in generality in assuming, as we will do in the rest of the paper without comment, that <u>every point on ∂G is essential</u>.

The paper is organized in the following fashion. In Chapter 2, we show that if π is a unital representation with domain $H^\infty(G)$, then π is unique provided that either S is a pure subnormal operator, or $\mu(\partial G)=0$. (A pure subnormal operator is one that has no nontrivial invariant subspace on which it is normal.)

In Chapter 3 we establish as a corollary to a more general result that any unital representation π is weak-star, weak-star continuous, provided that $S=\pi(\chi)$ is pure. In Chapter 4 we describe some spectral mapping theorems for a given unital representation. Our results describe (under various hypotheses) the spectrum of $\pi(f)$, denoted $\sigma(\pi(f))$, and the essential spectrum of $\pi(f)$, denoted $\sigma_e(\pi(f))$.

We have already discussed the outline of Chapter 5. This section illustrates how important the hypothesis of purity in the theorems of Chapters 2 and 3 are. As a corollary to our principal theorem, we show there are many unital representations π from $H^\infty(D)$ into $L^\infty(m)$. Throughout the paper D will denote the open unit disc and m will denote normalized Lebesgue measure on ∂D. Given a Blaschke product b whose zeros accumulate everywhere on ∂D and a function $f \in L^\infty(m)$ with $\|f\| \le 1$, we shall show there exists a unital representation π such that $\pi(b)=f$ (consult Example 40). This last fact then answers a uniqueness question found in [8]. The authors of this latter work ask: "If B is a polynomially bounded operator such that $B^k \to 0$ in the weak operator topology, can there exist two different norm-continuous representations each of which sends 1 to 1 and χ to B?" The answer is yes. If B is the bilateral shift (a normal operator), then there are many such representations. (On the other hand, if B is the unilateral shift (a pure subnormal operator), then there is only one unital representation.)

In Chapter 6 we investigate the question of whether, for a given region G, there is a probability measure μ on ∂G and a unital representation of $H^\infty(G)$ into $L^\infty(\mu)$ that is an isometry? The problem is completely solved for simply connected domains. A region of this last type supports an isometric unital representation if and only if it is nicely connected. The representations that are isometries are classified. What happens for an arbitrary domain? We do not know.

In Chapter 7 we define the notion of when two unital representations are partially subordinate and relate this concept to some issues of the earlier sections. In particular, we investigate the lattice structure of a representation and answer the question of when a representation is one-to-one. The material in the last chapter has already been discussed.

We have tried to keep the material in the last four chapters as self-contained as possible. (There are, however, times when we need to draw on some of the results in Chapters 2 and 3). Our primary reason for doing this, as indicated earlier, is that the problems addressed in these sections are (can be viewed as) purely function-theoretic ones; we hope that readers, who may not have an interest in the theory of subnormal operators, will still find some value in this material.

We close this chapter by asking a question that deals with an issue not addressed in this paper.

<u>Question</u> 2. If S is a pure subnormal operator acting on the Hilbert space $\mathcal{H}$ and π is an algebra homomorphism from $H^\infty(G)$ into $\mathcal{B}(\mathcal{H})$ with $\pi(1)=1$ and $\pi(\chi)=S$, then is π norm continuous?

If S has a cyclic vector, then S is unitarily equivalent to M_χ on $H^2(\mu)$ for some compactly supported measure in the plane. (The Hilbert space $H^2(\mu)$ consists of the closure of the polynomials in $L^2(\mu)$.) A result of Yoshino [41] states that $\{S\}' = \{M_\psi : \psi \in H^2(\mu) \cap L^\infty(\mu)\}$ if S has a cyclic vector. Hence, every operator belonging to $\{S\}'$ is subnormal.

If we assume S has a cyclic vector in Question 2, then the answer is yes. Recall the theorem that says any algebra homomorphism from one Banach algebra into a semi-simple commutative Banach algebra is continuous [12, Prop.4.2]. So to prove the fact, it suffices to show the only quasinilpotent operator in $\{S\}'$ is zero. If $\psi \epsilon H^2(\mu) \cap L^\infty(\mu)$ and M_ψ is quasinilpotent, then $\sigma(M_\psi) = \{0\}$. Hence, we see that $M_\psi = 0$ if we recall the fact that subnormal operators have the property that their spectral radius is equal to their norm. (Yes, the last argument did not use the assumption of purity.)

CHAPTER II

UNIQUENESS OF REPRESENTATIONS

Let S be a subnormal operator on $\mathcal{H}$ and let N be its minimal normal extension on $\mathcal{K}$. The spectral measure for N will be denoted by E. Suppose π_1 and π_2 are two unital representations from $H^\infty(G)$ into $\mathcal{B}(\mathcal{H})$ such that $\pi_i(\chi)=S$ for i=1,2. Note that in the next three lemmas and the corollary, we do not assume S is a pure operator (in particular, the case S=N is allowed).

Lemma 3. (Assume the notation in the preceding paragraph.) Let W be a bounded region containing G and let $g \epsilon H^\infty(W)$. Let $\lambda \epsilon W$ and let $\Delta(r) = \Delta(\lambda,r)$ be the open disc centered at λ with radius r. Then

$$\overline{\lim_{r\to 0}} \frac{\|E(\Delta(r))(\pi_1(g)-\pi_2(g))\|}{r^2} = 0.$$

Proof. Fix $g \epsilon H^\infty(W)$ and $\lambda \epsilon W$ and choose $r>0$ such that $\Delta(r) \subset W$. Let $h_\lambda(z)$ be the first three terms of the Taylor series expansion of g about λ; i.e.,

$$h_\lambda(z) = g(\lambda) + g'(\lambda)(z-\lambda) + \frac{g''(\lambda)(z-\lambda)^2}{2!}.$$

Then $g-h_\lambda = (\chi-\lambda)^3 q_\lambda$ where the supremum norm of q_λ on W, denoted $\|q_\lambda\|$, is bounded by M, where M depends only on some universal constants, $\|g\|$ and the distance of λ to ∂W. (To see this estimate on M use the Cauchy estimates on the derivatives of g at λ and the inequality

$$|q_\lambda| \le (|g| + |h_\lambda|)(|\chi-\lambda|)^{-3}.)$$

We then have

$$\|E(\Delta)\,[\pi_1(g)-\pi_2(g)]\|$$

$$\leq \|E(\Delta)\ [\pi_1(g)-h_\lambda(S)]\| + \|E(\Delta)[h_\lambda(S)-\pi_2(g)]\|$$

$$= \|E(\Delta)(S-\lambda)^3\pi_1(q_\lambda)\| + \|E(\Delta)(S-\lambda)^3\pi_2(q_\lambda)\|$$

$$\leq \|E(\Delta)(S-\lambda)^3\|[\|\pi_1(q_\lambda)\| + \|\pi_2(q_\lambda)\|]$$

$$\leq \|(N-\lambda)^3E(\Delta)\|\ M(\|\pi_1\| + \|\pi_2\|)$$

$$\leq \|(N-\lambda)^3E(\Delta)\|\ M(\|\pi_1\| + \|\pi_2\|)$$

$$\leq r^3\ M\ (\|\pi_1\| + \|\pi_2\|). \blacksquare$$

<u>Lemma</u> 4. With the same notation and assumptions as in Lemma 3, we have

$$\|E(W)\ (\pi_1(g)-\pi_2(g))\| = 0$$

for every $g \in H^\infty(W)$.

<u>Proof</u>. If not, then by the regularity of E there exists a compact set $K \subset W$ such that

$$\eta \equiv \|E(K)(\pi_1(g)-\pi_2(g)\| > 0.$$

Let d be the diameter of K. Construct a square that has sides of length d and that contains K. Partition the square into four congruent squares each of which has sides of length $d/2$. One of these four squares, say T_1, has the property that

$$\|E(T_1 \cap K)(\pi_1(g)-\pi_2(g))\| \geq \eta/4.$$

Continuing this process by induction, we construct a sequence $\{T_n\}$ of squares with the following properties:

$$T_{n+1} \subset T_n;$$

the sides of T_n have length $d/2^n$;

and

$$\|E(T_n \cap K)(\pi_1(g)-\pi_2(g))\| \geq \eta/4^n.$$

Let λ be such that $\{\lambda\} = \cap T_n$. Since K is compact and $T_n \cap K$ is nonempty for every n, it follows that $\lambda \in K$.

Let $r_n = d/2^{n-1}$. Then $T_n \subset \Delta(\lambda, r_n)$ and

$$\|E(\Delta(\lambda,r_n))(\pi_1(g)-\pi_2(g))\|$$

$$\geq \|E(T_n \cap K)(\pi_1(g)-\pi_2(g))\|$$

$$\geq \eta/4^n.$$

Thus,

$$\|E(\Delta(\lambda,r_n))(\pi_1(g)-\pi_2(g))\| \geq \eta r_n^2/4d^2.$$

Letting $n\to\infty$, we see that the conclusion of Lemma 3 is contradicted. ∎

In a way, the result of Lemma 4 is unsatisfactory. The lemma implies that $E(W)\pi_1(g)$ and $E(W)\pi_2(g)$ are equal; but, intuition tells one (based on many examples) that $\pi(g)$ should be multiplication by g on the space $E(W)\mathcal{H}$ for $g \in H^\infty(W)$. (We continue to use the setting of Lemmas 3 and 4.) Note, however, $E(W)\mathcal{H}$ may not be contained in $\mathcal{H}$ and, therefore, vectors in the subspace $E(W)\mathcal{H}$ are not necessarily in the domain of $\pi(g)$. However, by an appropriate orthogonal decomposition of the space $\mathcal{K}$, we can place our beliefs on solid ground.

Let μ be a scalar-valued spectral measure for N. Then [10, Chapter 2, Section 9] there exists a sequence (possibly finite) of measures $\{\mu_i\}$ such that $\mu=\mu_1$, $\mu_{i+1}<<\mu_i$ for all i and N is (unitarily equivalent to) the operator $\oplus_i M_\chi$ on the space $\oplus_i L^2(\mu_i)$. Assume that π is a unital representation defined on $H^\infty(G)$ and N is the minimal normal extension of $\pi(\chi)$.

It follows easily from the fact that π is a homomorphism that $\bar{G} \supset \sigma(S)$. Recalling the facts that $\sigma(S) \supset \sigma(N)$, that $\partial\sigma(S) \subset \partial\sigma(N)$ and that $\sigma(N) = \text{spt } \mu$, we see spt $\mu_i \subset \bar{G}$ for all i.

<u>Lemma</u> 5. Let π be a unital representation of $H^\infty(G)$ into $\mathcal{B}(\mathcal{H})$ where $\pi(\chi)=S$. Let W be a region containing G. Using a unitary operator, if necessary, we assume the minimal normal extension N of $\pi(\chi)$ is

$$\oplus_i M_\chi \text{ on } \oplus_i L^2(\mu_i).$$

(This decomposition is that described in the above.) Let $t = \oplus_i t_i$ be a vector in $\mathcal{H}$ and let $f \in H^\infty(W)$. Choose $\phi_i \in L^2(\mu_i)$ for each i so that

$$\pi(f)t = \oplus_i \phi_i$$

Then, for each i, we have

$$\phi_i = f t_i$$

almost everywhere $\mu_i|_W$.

The measure $\mu_i|_W$ is the restriction of μ_i to W. Clearly Lemma 5 implies Lemma 4. The proof of Lemma 5 relies on the following fact whose proof is left to the reader. (One way of proving this fact is to use the technique of the proof of Lemma 4.)

Fact 6. If μ is a positive compactly supported measure in the plane, then the set

$$\{w \in \mathbb{C}: \quad \int \frac{d\mu(z)}{|z-w|^2} < \infty\}$$

has μ-measure zero.

Proof of Lemma 5. Suppose to the contrary that there exist an i_0 and $\varepsilon > 0$ such that

$$|\phi_{i_0} - f t_{i_0}| \geq \varepsilon$$

on a compact subset $E \subset W$ with $\mu_{i_0}(E) > 0$. We may assume, without loss of generality, $|t_{i_0}| \leq M$ on E for some positive constant M. Fix a point $w_0 \in E$ such that $\mu_{i_0}(E \cap U) > 0$ for each neighborhood U of w_0.

By the continuity of f and the boundedness of t_{i_0}, there exists a neighborhood U of w_0, $U \subset W$ such that

$$|f(w) - f(w_0)| \; |t_{i_0}| < \varepsilon/2$$

for all $w \in U$. Observe that, for each $w \in U$,

$$(\pi(f - f(w)))t = \oplus_i (\phi_i - f(w)t_i).$$

For each $w \in U$, we define a function $h_w \in H^\infty(W)$ via

$$h_w(z) = \begin{cases} \frac{f(z)-f(w)}{z-w} & z \in G \setminus \{w\} \\ f'(w) & z=w. \end{cases}$$

We compute:

$$\pi(f-f(w)) = \pi((\chi-w)h_w)$$

$$= (S-w)\pi(h_w)$$

$$= (N-w)\pi(h_w)$$

$$= ((\sum_i M_\chi)-w)\pi(h_w).$$

Hence, for each $w \in U$, the function

$$\bigoplus_i \frac{(\phi_i - f(w)t_i)}{\chi - w} = \pi(h_w)t$$

belongs to $\bigoplus_i L^2(\mu_i)$. But, everywhere on $E \cap U$, we have

$$|\phi_{i_0} - f(w)t_{i_0}| \geq |\phi_{i_0} - ft_{i_0}| - |f-f(w)|\,|t_{i_0}|$$

$$\geq \varepsilon - \varepsilon/2$$

$$= \varepsilon/2.$$

Hence, it follows that $\frac{1}{\chi - w} \epsilon\ L^2(\mu_{i_0}|_{U \cap E})$ for every $w \epsilon U$. Applying Fact 6, we get that $\mu_{i_0}(U \cap E)=0$; a clear contradiction to the fact $\mu_{i_0}(U \cap E)>0$. ■

A uniqueness theorem for unital representations now follows for a wide variety of examples.

Corollary 7. Let S be a subnormal operator and let N be its minimal normal extension. If π_i for i=1,2 are two unital representations defined on $H^\infty(G)$ with $\pi_i(\chi)=S$, then $\pi_1=\pi_2$ provided that $E(\partial G) = 0$ where E is the spectral measure for N.

We see, from this corollary, that the representation defined in Example 1 is unique. As mentioned in the introduction (see Chapter 5) this uniqueness can fail if $E(\partial G) \neq 0$. But not if $\pi(\chi)$ is pure. Before we prove this last statement, we present two results from the theory of function algebras. They are well-known to the workers in this area, but we include sketches of the proofs for completeness. (The reader who wants to fill in the gaps may consult [18,19].)

Lemma 8. Let G be a bounded region and $\lambda \in \partial G$. Let $r>0$ and small enough so that $G \cap (\mathbb{C} \setminus \Delta(\lambda,r)) \neq \phi$. If $g \in H^{\infty}(G)$, then there exist functions $g_i \in H^{\infty}(W_i)$ for $i=1,2$ such that $g=g_1+g_2$ where the regions W_i are given by

$$W_1 = G \cup (\mathbb{C} \setminus \Delta(\lambda,r)^{-})$$

and

$$W_2 = G \cup \Delta(\lambda,r/2).$$

Proof. Let g be defined on all of $\mathbb{C}$ by setting $g=0$ on $\mathbb{C} \setminus G$. Let ϕ be smooth function (continuous partial derivatives) defined on $\mathbb{C}$ with compact support contained in $\Delta(\lambda,r)$, such that $\phi=1$ on $\Delta(\lambda,r/2)$ and $\phi=0$ outside $\Delta(\lambda,r)$.

We define g_1 on $\mathbb{C}$ as follows:

$$g_1(\lambda) \equiv g(\lambda)\phi(\lambda) + (1/\pi)\iint \frac{g(z)}{z-\lambda} \frac{\partial \phi}{\partial \bar{z}} dxdy$$

(where $z=x+iy$). It follows (consult [19, p.222]) that, in the sense of distributions,

$$\frac{\partial g_1}{\partial \bar{z}} = \phi \frac{\partial g}{\partial \bar{z}}.$$

Therefore, g_1 is analytic on $G \cup (\mathbb{C} \setminus \Delta(\lambda,r))$. It is a bounded function on all of $\mathbb{C}$ because the integral used in defining g_1 is a continuous function of λ on $\mathbb{C}$ that vanishes at ∞ (being the convolution of the locally integrable function $1/\chi$ with a bounded function with compact support).

If we define g_2 as $g_2 = g-g_1$, then it follows that

$$\frac{\partial g_2}{\partial \bar{z}} = (1-\phi)\frac{\partial g}{\partial \bar{z}} \quad ;$$

and the conclusion of the lemma follows.■

If K is a compact subset of the plane, let R(K) be the uniform closure of the rational functions with poles off K. This space R(K) is a function algebra contained in C(K), the

Banach algebra of continuous functions on K.

Lemma 9. Let G be a bounded region contained in $\mathbb{C}$. Then

$R(\partial G) = C(\partial G)$.

Proof. For any compact set K in the plane $R(K) = C(K)$ if and only if every point of K is a peak point for R(K) (see Bishop's theorem [19,p.54]). Furthermore, if U is a component of $\mathbb{C} \setminus K$, then every point λ on the boundary of U is a peak point of R(K) (consult [19,Corollary 4.3, p. 205]). The result of the lemma immediately follows if we note that G is a component of $\mathbb{C} \setminus \partial G$. ∎

We are now ready to prove the main theorem of this section.

Theorem 10. Let π_i for i=1,2 be two unital representations of $H^\infty(G)$ with $\pi_1(\chi) = \pi_2(\chi) = S$. If S is a pure subnormal operator, then $\pi_1 = \pi_2$.

Proof. Choose a point $\lambda \in \partial G$, and let $g \in H^\infty(G)$. Using this point λ in Lemma 8, we can write $g = g_1 + g_2$ where each g_i is a bounded analytic function in a region $W_i \supset G$ and $W_i \cap (\mathbb{C} \setminus \bar{G}) \neq \phi$ for i=1,2. To show $\pi_1(g) = \pi_2(g)$, it suffices to show $\pi_1(g_i) = \pi_2(g_i)$ for i=1,2.

Fix an element $x \in \mathcal{H}$, and let $y = (\pi_1(g_1) - \pi_2(g_1))x$. Clearly $y \in \mathcal{H}$ and, by Lemma 4, we have $y = E(\partial G \setminus W_1)y$. (Recall that $\bar{G} \supset \sigma(S) \supset \sigma(N)$; hence it follows that $E(\mathbb{C} \setminus \bar{G}) = 0$.) We want to show y=0.

Let $\tilde{\mathcal{H}}$ be the closure of the linear manifold {r(s)y: r is a rational function with poles off $\sigma(S)$}. It follows then that $\tilde{\mathcal{H}} \subset \mathcal{H}$, and that $\tilde{\mathcal{H}}$ is a closed invariant subspace for r(S) where r is any rational function with poles off $\sigma(S)$. Let $\tilde{S}$ be the subnormal operator obtained by restricting S to $\tilde{\mathcal{H}}$. Clearly, $\sigma(\tilde{S}) \subset \bar{G}$.

We will show that $\sigma(\tilde{S}) \subset \partial G$. It will then follow from Lemma 9 and [10,p.302] that $\tilde{S}$ is a normal operator. Since S is pure, this implies $\tilde{\mathcal{H}} = (0)$; hence y=0. A similar argument shows $\pi_1(g_2) = \pi_2(g_2)$; the proof of the theorem will then be completed.

Let $\widetilde{\mathcal{K}}$ be the closure of the linear manifoild $\{\sum_{k=1}^{n} N^{*k} w_k$: n a nonnegative integer, $w_k \in \widetilde{\mathcal{H}}\}$. It is easy to show that the operator $\widetilde{N}$ obtained by restricting N to $\widetilde{\mathcal{K}}$ is the minimal normal extension of $\widetilde{S}$. Furthermore, we have that $\sigma(\widetilde{N}) \subset \partial G \setminus W_1$ because of the fact that $E(W_1)g=0$ for all $g \in \widetilde{\mathcal{K}}$. Hence, $\partial\sigma(\widetilde{S}) \subset \partial G \setminus W_1$. (Again recall the fact, if A is a subnormal operator and B is its minimal normal extension, then $\sigma(A)$ is the union of $\sigma(B)$ with some of its holes; ;consult [10,p.131].)

Now, using the properties of W_1, we can choose an open disc $\Delta \subset W_1$ such that $\Delta \cap (\mathbb{C} \setminus \overline{G}) \neq \phi$ and $\Delta \cap G \neq \phi$. Since $\sigma(\widetilde{S}) \subset \overline{G}$ and $\partial\sigma(\widetilde{S}) \cap W_1 = \phi$, we have that $\sigma(\widetilde{S}) \cap \Delta = \phi$.

Let $\gamma \in \Delta \cap G$. If β is any other point of G, we can joint β to γ by a polygonal line segment, Γ, that lies entirely within G. Since $\gamma \notin \sigma(\widetilde{S})$ and $\partial\sigma(\widetilde{S}) \subset \partial G$, an easy compactness argument yields $\Gamma \cap \sigma(\widetilde{S}) = \phi$. Hence $\beta \notin \sigma(\widetilde{S})$. Since β was an arbitrary point of G we have $G \cap \sigma(\widetilde{S}) = \phi$. Hence, $\sigma(\widetilde{S}) \subset \partial G$ as we promised to show.∎

Recall, again, the fact that ran $\pi \subset \{S\}'$ for a unital representation. A much stronger result holds, if S is pure. Recall that, for any $T \in \mathcal{B}(\mathcal{H})$,

$$\{T\}'' \equiv \{R \in \mathcal{B}(\mathcal{H}):\ RA=AR \text{ for all } A \in \{T\}'\}.$$

<u>Corollary</u> 11. Let π be a unital representation defined on $H^\infty(G)$. If $S=\pi(\chi)$ is pure or $E(\partial G)=0$, where E is the spectral measure for the minimal normal extension of S, then ran $\pi \subset \{S\}''$.

<u>Proof</u>. Fix an invertible operator $T \in \{S\}'$ and define a (new) unital representation π_1 via

$$\pi_1(g) \equiv T^{-1}\pi(g)T$$

for all $g \in H^\infty(G)$. (One easily checks that π_1 is a continuous representation, $\pi_1(1)=1$ and $\pi_1(\chi)=S$.) From Theorem 10 or Corollary 7, whichever case is applicable, we see that $\pi_1 = \pi$. Hence, for any $g \in H^\infty(G)$, the operator $\pi(g)$ commutes with all the invertible operators in $\{S\}'$. Clearly, the conclusion of the corollary now follows.∎

In Section 5 we will exhibit an example (Example 41) of a unital representation π where $\pi(\chi)$ equals the bilateral shift of multiplicity two and ran $\pi \not\supset \{\pi(\chi)\}''$. We close this section with a conjecture and a question.

Conjecture 12. Let π be a unital representation defined on $H^\infty(G)$. If $S=\pi(\chi)$ is pure, then $\pi(f)$ is a subnormal operator for every $f \in H^\infty(G)$.

As noted earlier, if S has a cyclic vector, then $\{S\}'$ consists of only subnormal operators. The conjecture is trivally true in this case. Note in this setting, we can construct a unital representation $\tilde{\pi}$ defined on $H^\infty(G)$ into $\mathcal{B}(\mathcal{K})$, where $\mathcal{K}$ is the Hilbert space on which the minimal normal extension acts, such that $\tilde{\pi}(f)$ is normal and

$$\pi(f) = \tilde{\pi}(f)|_{\mathcal{H}}$$

for all $f \in H^\infty(G)$.

Question 13.1. Is an extension $\tilde{\pi}$ of π, like that constructed in the preceding paragraph, always possible (even if S does not have a cyclic vector)?

We do not know of an example of a unital representation where the conjecture fails even if $\pi(\chi)$ is not pure.

Remark 13.2. Although the emphasis of this work is focused on subnormal operators and their representations, the techniques developed are sufficiently general to handle other situations. As an illustration, let T be a completely non-unitary contraction on a Hilbert space $\mathcal{H}$ and let Φ_T denote the algebra homomorphism of $H^\infty(D)$ into $\mathcal{B}(\mathcal{H})$ described in[30, Chapter III, Theorem 2.1]. (We refer the reader to [42] for an accounting of how this particular homomorphism has been used to push the "Scott Brown process" into the realm of other operators.) The following theorem along with the example presented after it answers some questions suggested to us by C. Foias (private communication).

Theorem 13.3. Let T be a contraction on $\mathcal{H}$ that either belongs to class $C_{0.}$ or class $C_{.0}$, (consult [30, p.72]). If π is continuous algebra homomorphism of $H^\infty(D)$ into $\mathcal{B}(\mathcal{H})$ that sends the identity to the identity and χ to T, then

$$\pi = \Phi_T.$$

(For contractions in either of those two classes this theorem is a generalization of [30,Remark 1,p.114].)

Proof. If $T \in C_{.0}$, then we consider the homomorphism $\pi^*: H^\infty(D) \to \mathcal{B}(\mathcal{H})$ defined in the beginning of the next chapter that sends χ to $T^* \in C_{0.}$. If the theorem is valid for operators in this latter class, then $\pi^* = \Phi_{T^*}$; hence, $\pi = \Phi_T$. So we assume that $T \in C_{0.}$.

Fix a vector $x \in \mathcal{H}$ and a function $f = \sum_{k=0}^{\infty} a_k \chi^k$ in $H^\infty(D)$ with norm less than or equal to one. We need to show that

$$\pi(f)x = \Phi_T(f)x.$$

Let $\{\sigma_n\}_{n=1}^{\infty}$ denote the sequence of Cesàro means of f. It is well-known [28,Chapter II] that

$$\|\sigma_n\| \le \|f\|$$

(because you recover σ_n by integrating f up against the Fejer's kernel), and that

$$\sigma_n \to f \text{ weak-star in } L^\infty(m).$$

Each σ_n is a polynomial of degree n−1, so we may write

$$\sigma_n = \sum_{h=0}^{n-1} C_{k,n} \chi^k.$$

Because the mapping $g \to g^{(k)}(0)$ is a weak-star continuous linear functional on $H^\infty(m)$ for each k, we see that for each k

$$a_k = \lim_{n \to \infty} C_{k,n}.$$

Now fix $\varepsilon > 0$ and N a positive integer. From the last equality we may choose an integer M=M(N) such that

$$|a_k - C_{k,M}| < \frac{\varepsilon}{N}$$

for all k=0,1,2,...,N. Fixing this M, we now let

$$P_N(z) = \sum_{k=0}^{N} (a_k - C_{k,M}) z^k.$$

Obviously from our choice of σ_M, we have that

$$\|p_N\| \leq \epsilon,$$

and that the first N Fourier coefficients of the function τ defined by

$$\tau = f - \sigma_M - p_N$$

are zero.

Hence, we see that

$$\|\tau\| \leq 2+\epsilon$$

and that there exists $h \epsilon H^\infty$ such that

$$\tau = \chi^N h.$$

From our estimate on the norm of τ we have $\|h\| \leq 3$. Thus, we have (the pointwise idea of the proof of Lemma 3)

$$\begin{aligned} \|\pi(f)x - \Phi_T(f)x\| &= \|\pi(\tau)x - \Phi_T(\tau)x\| \\ &\leq \|\pi(h) - \Phi_T(h)\| \ \|T^N x\| \\ &\leq 3(\|\pi\| + \|\Phi_T\|) \ \|T^N x\|. \end{aligned}$$

It obviously follows that

$$\pi(f)x = \Phi_T(f)x, \cdot$$

since N was arbitrary and $T \epsilon C_{0.}$. ∎

One might inquire if the result of Theorem 13.3 holds for all completely non-unitary contractions, or the (canonical class of emphasis in [42]) class of BCP completely non-unitary contractions. The answer is no.

<u>Example</u> 13.4. Let W be the bilateral shift of multiplicity one acting on ℓ^2 with the canonical basis $\{e_n\}_{n=-\infty}^{\infty}$. Let $\{w_n\}_{n=-\infty}^{\infty}$ be a sequence of positive numbers and define S on ℓ^2 by requiring that for all n we have

$$Se_n = w_n e_n.$$

It is a routine exercise to show that the w_n's can be chosen such that S is invertible and that

$$C \equiv SWS^{-1},$$

a weighted bilateral shift, is a strict contraction ($\|Cx\| < \|x\|$ for all $x \in \ell^2$); hence, C is completely non-unitary.

In Chapter V, Example 40 we will construct two distinct representations π_i of $H^\infty(D)$ into $\mathcal{B}(\ell^2)$ satisfying $\pi_i(1)=1$ and $\pi_i(\chi)=W$. It is routine to verify that the following equation,

$$\eta_i \equiv S\pi_i S^{-1},$$

then defines two distinct representations on $H^\infty(D)$ with

$$\eta_i(1)=1 \text{ and } \eta_i(\chi)=C.$$

Finally, if A is any completely non-unitary BCP contraction (on ℓ^2), then

$$T \equiv A \oplus C$$

shares the same properties that A does and the formula

$$\tau_i \equiv \Phi_A \oplus \eta_i$$

defines two distinct representations on $H^\infty(D)$ with $\tau_i(1)=1$ and $\tau_i(\chi) = T$.

We would like to emphasize before concluding this chapter that the last result, Theorem 13.3, does not generalize the earlier results in the chapter or the results of the next one (the function spaces $H^\infty(D)$ and $H^\infty(G)$ for G an arbitrary bounded domain are vastly different).

CHAPTER III

CONTINUITY PROPERTIES OF UNITAL REPRESENTATIONS

Throughout this section we will assume π is a unital representation from $H^\infty(G)$ into

$\mathcal{B}(\mathcal{H})$ with $\pi(\chi)$ (=S) equal to a pure subnormal operator. The spectral measure for N, the minimal normal extension of S, will be denoted by E. Recall from the introduction, $H^\infty(G)$ is a weak-star closed subalgebra of $L^\infty(G)$, the dual of $L^1(G)$, and a sequence $\{f_n\}$ in $H^\infty(G)$ converges weak-star to f if and only if the sequence is bounded and converges to f pointwise everywhere on G.

Let $G^* = \{\lambda \in \mathbb{C}: \bar{\lambda} \in G\}$. If $f \in H^\infty(G^*)$, then we define a function $f^\# \in H^\infty(G)$ via the equation

$$f^\#(z) = \overline{f(\bar{z})}$$

for all $z \in G$. The mapping $f \to f^\#$ is an isometric weak-star homeomorphism of $H^\infty(G^*)$ onto $H^\infty(G)$. From the unital representation π, we define another unital continuous homomorphism π^* of $H^\infty(G^*)$ into $\mathcal{B}(\mathcal{H})$ by defining

$$\pi^*(f) \equiv (\pi(f^\#))^*$$

for all $f \in H^\infty(G^*)$; note, $\pi^*(\chi) = S^*$. (The only reason in defining the domain of π^* to be $H^\infty(G^*)$, instead of $H^\infty(G)$, is that π^* becomes homogeneous. If one is willing to live with conjugate linear representations, one should define π^* on $H^\infty(G)$ in the obvious way; $\pi^*(f)=(\pi(f))^*$.) The main theorem of this section is the following.

<u>Theorem</u> 14. With the definitions stated above, the representation π^* is weak-star, s.o.t. sequentially continuous. That is, if $\{f_n\}$ is a sequence in $H^\infty(G^*)$ that converges weak-star, then $\{\pi^*(f_n)\}$ converges in $\mathcal{B}(\mathcal{H})$ in the strong operator topology, s.o.t.

This theorem deserves to be a corollary of the nice results obtained in [2]. The proof relies heavily on the techniques given there. For completeness, we will give the results used from this work. Before we do this, let us point out two corollaries to Theorem 14.

<u>Corollary</u> 15. Let S be a pure subnormal operator with $\sigma(S) \subset \bar{D}$. Then $(S^*)^n \to 0$ s.o.t. as $n \to \infty$. Using the terminology of [30,p.72], we see that every pure subnormal contraction

is a $C_{.0}$ contraction.

Proof. Using [11], we see there exists a unital representation from $H^\infty(D)$ into $\mathcal{B}(\mathcal{H})$ with $\pi(\chi)=S$. The result now follows from Theorm 14, if we observe that $D^*=D$ and $\{\chi^n\}$ converges to zero weak-star.∎

Corollary 16. The unital representation π is weak-star continuous (i.e., $\{\pi(f_\alpha)\}$ converges to $\pi(f)$ in $\mathcal{B}(\mathcal{H})$ with the weak-star topology if the net $\{f_\alpha\}$ converges to f weak-star in $H^\infty(G)$).

Proof. It suffices to show that if $\{f_n\}$ is a sequence in $H^\infty(G)$ that converges weak-star to zero, then $\{\pi(f_n)\}$ converges to zero in $\mathcal{B}(\mathcal{H})$ with the weak-operator topology. (Reasons: $H^\infty(G)$ is the dual of a separable Banach space, and the weak operator topology and the weak-star topology on $\mathcal{B}(\mathcal{H})$ agree on balls.)

Observing that $\{f_n^\#\}$ converges weak-star to zero in $H^\infty(G^*)$, we obtain the result from Theorem 14 and the following estimate for any two vectors x and y in $\mathcal{H}$:

$$|<\pi(f_n)x,y>| = |<x,\pi(f_n)^* y>|$$

$$= |<x,\pi^*(f_n^\#)y>|$$

$$\leq \|x\| \quad \|\pi^*(f_n^\#)y\|. ∎$$

Proof of Theorem 14. Suppose to the contrary that there exist a sequence of functions $\{f_n\}$ belonging to $H^\infty(G)$ with $\|f_n\|\leq 1$ for all n and $f_n\to 0$ pointwise on G, and a vector $x\in\mathcal{H}$ with $\|x\|=1$, such that

$$\|\pi(f_n)^* x\|\to a$$

where $a>0$. If $x_n \equiv \pi(f_n)^* x$, then, by dropping to a subsequence, if need be, we can assume that $\pi(f_n)x_n$ converges weakly to a vector $y\in\mathcal{H}$. Observe $y\neq 0$ because

$$<y,x> = \lim_{n\to\infty} <\pi(f_n)x_n,x>$$

$$= \lim_{n\to\infty} \langle x_n, \pi(f_n)^* x \rangle$$

$$= a^2.$$

Using the proof of Theorem 2.1 in [2], we construct a well-defined bounded linear map Γ from $R(\partial G)$ into $\mathcal{H}$ satisfying

(17) $q(S)\Gamma(r) = p(S)y$

for any rational function $r = p/q$ with poles off ∂G. (Here p and q are polynomials. The reader should check [2] for the details on the construction of Γ.) From Lemma 9, we see the domain of Γ is $C(\partial G)$. We now shall establish

(18) $$(S-\lambda)\Gamma(f) = \Gamma((\chi-\lambda)f)$$

for all $\lambda \in \mathbb{C}$ and all $f \in C(\partial G)$. From Equation 17, we have, for any polynomials p and q,

$$q(S)\Gamma((\chi-\lambda)p/q) = (S-\lambda)p(S)y$$

$$= (S-\lambda)q(S)\Gamma(p/q)$$

$$= q(S)\ ((S-\lambda)\Gamma(p/q)).$$

Since S is a pure subnormal operator, it follows from [31] that q(S) is pure too; hence q(S) has no eigenvalues (cf. [10, p.141]). From this latter fact and the last equation, we see Equation 18 is established for every rational function with poles off ∂G. The result now follows from Lemma 9 and the continuity of Γ.

Thus, (from Equation 18) we see that, for all $\lambda \in G$ and all $f \in C(\partial G)$,

$$\Gamma(f) = (N-\lambda)\Gamma(f/(\chi-\lambda)).$$

Now recall that if F is a Borel subset of $\mathbb{C}$ and if

$$x \in \bigcap_{\lambda \in F} \operatorname{ran}(N-\lambda),$$

then $E(F)x = 0$ (see Fact 6). Combining both facts in this paragraph, we see that

$$\Gamma(f) \in E(\partial G)\mathcal{K} \cap \mathcal{H}$$

for every $f \in C(\partial G)$; in particular, $y = \Gamma(1)$ belongs to $E(\partial G)\ \mathcal{K}\cap\mathcal{H}$.

If q is any polynomial with no zeros on ∂G, then $q(N)|_{E(\partial G)\mathcal{K}}$, which we identify with $q(NE(\partial G))$ in the obvious way, is an invertible operator. Using Equation 17, we have, for every rational function p/q with poles off ∂G,

$$(p/q)(NE(\partial G))y = (q(NE(\partial G)))^{-1}[q(N)\Gamma(p/q)]$$
$$= \Gamma(p/q).$$

Extending this last equality to all $f \in C(\partial G)$ by Lemma 9, we get $\Gamma(f) = f(N)y$ for every $f \in C(\partial G)$. It follows then that the closure of the linear manifold $\{\Gamma(f): f \in C(\partial G)\}$ is a nonzero invariant subspace for S on which it is normal; a contradiction to the hypothesis that S is pure.■

Remark 19. If the answer to Question 13 has an affirmative answer, then this extension $\tilde{\pi}$ is weak-star continuous. This follows from Corollary 16 and [10, Theorem 12.9, p. 208].

Remark 20.1. Combining Corollary 16 with [30, Proposition 2.1, p. 245], we see that the characteristic function $\Theta_S(\lambda)$ of every pure subnormal contraction is an inner function. It seems natural to ask: "What inner characteristic functions arise from the pure subnormal contractions?"

Remark 20.2. This chapter has pointed out some consequences of the techniques in [2] to the theory of pure subnormal operators. The results in [2], however are very general and what is presented here generalizes to much larger classes of operators. In fact, the reader familiar with [2], in particular the remarks at the end of section 3 there, should realize that what makes the arguments of this chapter work follow from the fact: If S is a pure subnormal operator and $\mathcal{M}$ is an invariant subspace for S, then

$$R(\sigma(S\,|_{\mathcal{M}})) \neq C(\sigma(S\,|_{\mathcal{M}})).$$

Consequently, a pure subnormal operator has no eigenvalues, and, in the setting given at the beginning of this chapter, S has no invariant subspace $\mathcal{M}$ such that

$$\sigma(S\,|_{\mathcal{M}}) \subseteq \partial G.$$

Any operator T enjoying these last two properties will share the same structure set out in this chapter that a pure subnormal operator does. In particular then, if T is also a completely non-unitary contraction and Φ_T is the algebra homomorphism described in Remark 13.2, then Φ_T^* is weak-star, s.o.t. sequentially continuous.

SUBNORMAL OPERATORS

CHAPTER IV

SPECTRAL MAPPING THEOREMS

Every unital representation π (defined on $H^\infty(G)$) defines a functional calculus for S. It is natural to inquire about the relationship between $\sigma(\pi(f))$(or $\sigma_e(\pi(f))$) and the values of f on G. In this section we yield to this compulsion and describe some partial results on these two problems. First we study the problem of describing $\sigma(\pi(f))$ in terms of values of f; we then turn to the description of $\sigma_e(\pi(f))$.

There is a natural inclusion between the values of f on $G \cap \sigma(S)$ and $\sigma(\pi(f))$. (Recall, $\sigma(S) \subset \bar{G}$ because π is an algebra homomorphism.)

<u>Lemma</u> 21. Let π be a unital representation defined on $H^\infty(G)$. If $S = \pi(\chi)$, then

$$\sigma(\pi(f)) \supset f(G \cap \sigma(S))^-$$

for every $f \in H^\infty(G)$.

<u>Proof</u>. Let $\lambda \in G \cap \sigma(S)$. Either $S-\lambda$ has closed range that is not all of $\mathcal{H}$($S \in \mathcal{B}(\mathcal{H})$), or $S-\lambda$ is not bounded below. Fix $f \in H^\infty(G)$ and let $g \in H^\infty(G)$ be chosen such that

$$f-f(\lambda) = (\chi-\lambda)g;$$

hence,

$$\pi(f)-f(\lambda) = (S-\lambda)\pi(g).$$

It is obvious then that either $\pi(f)-f(\lambda)$ has closed range that is not all of $\mathcal{H}$, or $\pi(f)-f(\lambda)$ is not bounded below.∎

The reverse inclusion is easy to establish if $\sigma(S)$ is "big" in G.

<u>Corollary 22</u>. Let π be a unital representation defined on $H^\infty(G)$. If $S = \pi(\chi)$ and $\bar{G} \subset \sigma(S)$, then

$$\sigma(\pi(f)) = f(G)^-$$

Proof. Since π is an algebra homomorphism, $\pi(f^{-1}) = (\pi(f))^{-1}$ for every invertible function f in $H^\infty(G)$. Therefore, if $0 \notin f(G)^-$, then $\pi(f)$ is invertible. Hence, $f(G)^- \supset \sigma(\pi(f))$ for every $f \in H^\infty(G)$. Now apply Lemma 21 to finish the proof.■

In passing, we note that the hypothesis of subnormality was not used in the proofs of the last lemma and corollary. Corollary 12 covers a wide variety of examples; but, of course, many interesting and important cases are left. Consider the following possibility. Let $G = \{z \in D: |z-(1/2)| > 1/2\}$. If

$$f(z) = \exp\ ((z+1)(z-1)^{-1})$$

for all $z \in D$,

$$S = M_\chi \text{ on } H^2(G),$$

and π: $H^\infty(D) \to \mathcal{B}(H^2(G))$ is the unital representation described in [11], then $0 \in f(D)^-$, but $\pi(f)^{-1}$ exists because $|f(z)| \geq e^{-1}$ for all $z \in G$. (Note, S is a pure subnormal operator.)

In investigating the reverse inclusion of Lemma 21, one should add at the onset the hypothesis of purity to S. There are many reasons for this. (The most obvious one being that, it then follows from Lemma 9 that $G \cap \sigma(S)$ is not vacuous.)

Recall that a set $F \subset G$ is a dominating set for $H^\infty(G)$ if

$$\sup_{z \in F} |f(z)| = \|f\| \equiv \sup_{z \in G} |f(z)|$$

for every $f \in H^\infty(G)$. Before we continue (and give our results), let us state what we believe is the general theorem in this context. (This conjecture with an affirmative answer would be a major generalization of the spectral mapping theorems given here and elsewhere.)

Conjecture 23. Assume π is a unital representation defined on $H^\infty(G)$. Assume further that,

S is pure,

and

$\sigma(S) \cap G$ is a dominating set for $H^\infty(G)$.

Then, for each $f \in H^\infty(G)$,

$\sigma(\pi(f)) = f(G \cap \sigma(S))^-$.

The conjecture is true if G=D (cf. [11]). We have been able to prove it in many other settings under the hypothesis, "the boundary of G is not too wild". Here is a representative example of these results.

Theorem 24. Assume π is a unital representation defined on $H^\infty(G)$ where S is a pure subnormal operator. Let $K = \bar{G}$. Assume further that

(i) $\operatorname{int} K = G$,

(ii) $\sigma(S) \cap G$ is a dominating set for $H^\infty(G)$,

(iii) There is a σ-curvilinear null set E such that for each $z \in (\partial K) \setminus E$

$$\liminf_{\sigma \to 0} \frac{\gamma(\Delta(z,\sigma) \setminus K)}{\sigma} > 0.$$

(Explanation of terminology and notion will come shortly.)

With these assumptions the following spectral mapping property holds for every $f \in H^\infty(G)$:

$\sigma(\pi(f)) = f(G \cap \sigma(S))^-$.

If F is a subset of a topological space, int F denotes the interior of F. Let us mention a few topological and measure theoretic criteria that imply condition (iii) of Theorem 24. If the diameters of the components of $\mathbb{C} \setminus K$ are bounded away from zero, then (iii) holds. If the inner boundary of K is countable, then (iii) holds. (The inner boundary of K consists of those points on ∂K that are not on the boundary of any component of $\mathbb{C} \setminus K$.) If the set of points on the boundary of K which are points of zero lower area density for $\mathbb{C} \setminus K$ is countable, then (iii) holds. The proof of these results follows from [23, Theorem 4; 24, Theorem 8.9] and the given estimates on analytic capacity in [19, Chapter VIII]. These results are well-known to the experts in this area and we omit the details. However, for completeness of the proof of Theorem 13, we give the definition of analytic capacity and some other relevant terminology.

For any set $E \subset \mathbb{C}$, the analytic capacity of E, denoted $\gamma(E)$, is defined as follows:

$\gamma(E) = \sup\{ |f'(\infty)| :$ f is a Borel function, bounded by one, analytic off a compact subset of E, and vanishing at ∞}.

A curvilinear null set is a subset of zero outer length lying on a twice continuously differentiable curve. A σ-curvilinear null set is a countable union of curvilinear null sets. A main part of the proof of Theorem 24 relies on the following theorem (cf. [23, Theorem 4 or 24, Theorem 8.9]).

Theorem 25 (Gamelin-Garnett). The following are equivalent for a compact set K:

(iv) $R(\partial K)=C(\partial K)$, and R(K) is pointwise boundedly dense in $H^\infty(\text{int } K)$.

(v) $\gamma(\Delta \setminus K) = \gamma(\Delta \setminus \text{int} K)$ for each bounded open set Δ.

(vi) There is a σ-curvilinear null set E

such that, for each $z \in (\partial K)\setminus E$, there exists $r \geq 1$ satisfying

$$\liminf_{\delta \to 0} \frac{\gamma(\Delta(z, r\delta)\setminus K)}{\gamma(\Delta(z,\delta)\cap \partial \text{int} K)} > 0.$$

Proof of Theorem 24. Using the facts that γ is monotone increasing and the capacity of a disc is its radius, we see hypothesis (iii) of our theorem implies condition (vi) of Theorem 25. We also note that R(K)=A(K), the uniform algebra of continuous functions in K that are analytic on int K (cf. [23, Theorem 2 or 24, Theorem 4.1]). An application of Theorem 1 in [14] then yields the fact that R(K) is strongly pointwise boundedly dense in $H^\infty(G)$; i.e., for every $g \in H^\infty(G)$, there exists a sequence $\{g_n\}$ in R(K) with $\|g_n\| \leq \|g\|$, such that $g_n \to g$ everywhere on G.

Let μ denote the scalar spectral measure for N (the minimal normal extension of S). If F is a compact set containing the support of μ, then $R^\infty(F,\mu)$ denotes the weak-star closure of R(F) in $L^\infty(\mu)$. If F also contains (the larger set) $\sigma(S)$, then $R^\infty(F,S)$ represents the weak-star closure of the set

$$\{r(S) : r \in R(F)\}$$

in $\mathcal{B}(\mathcal{H})$. Using the same techniques as in [10, pp. 207-209], one sees the map ρ from $R^\infty(F,\mu)$ onto $R^\infty(F,S)$ defined by

$$\rho(f) \equiv f(N)\,|\,_{\mathcal{H}}$$

is a unital representation that is an isometric-isomorphism and a weak-star homeomorphism.

It follows then, from hypothesis (iii) and Corollary 16, that $\pi(H^\infty(G))\subset R^\infty(K,S)$. In particular, $\pi(g)$ is subnormal for each $g\epsilon H^\infty(G)$. Hence π itself is an isometry. To see this, note first that $\|\pi(g)\|\geq\|g\|_\infty$ by hypotheses (ii) and Lemma 21. Furthermore, if $|\lambda|>\|g\|_\infty$, then $(g-\lambda)^{-1}\epsilon H^\infty(G)$. Hence $(\pi(g)-\lambda)^{-1}$ exists, and we conclude that

$$\sigma(\pi(g))\subset\{|z|\leq\|g\|_\infty\}.$$

Hence, $\|\pi(g)\|\leq\|g\|_\infty$ because the norm of a subnormal operator equals its spectral radius.

It now follows from the Krein-Smulian theorem (cf. [10, Proposition 6.7 in Chapter I]) that $H^\infty(G)$ and $\pi(H^\infty(G))$ are isometrically isomorphic and weak-star homeomorphic. We define a continuous algebra homomorphism Λ from $H^\infty(G)$ into $R^\infty(K,\mu)$ such that

$$\pi(g) = \rho_K(\Lambda(g))$$

for every $g\epsilon H^\infty(G)$. Note that $H^\infty(G)$ and $\Lambda(H^\infty(G))$ are isometrically isomorphic and weak-star homeomorphic. We now proceed with proof of the spectral equality. (The outline of this proof revolves around that one used in [11, Lemma 8.9].)

Fix $f\epsilon H^\infty(G)$ that is not constant; it suffices to show, from Lemma 21, that $\pi(f)$ is invertible if $0\notin f(G\cap\sigma(S))^-$. Let $\varepsilon>0$ be chosen so that $|f(\lambda)|\geq2\varepsilon$ for all $\lambda\epsilon G\cap\sigma(S)$. Let $H\equiv\{z\epsilon G\colon\ |f(z)|>\varepsilon\}$ and then set $F=\bar{H}$. It is easy to show that $\partial F\supset\partial K$, $\mathrm{int}F = H$, and if $\lambda\epsilon\partial F\cap G$, then $|f(\lambda)|=\varepsilon$.

We claim that if $\lambda\epsilon\partial F\cap G$ and $f'(\lambda)\neq0$, then λ is not on the inner boundary of F. To see this, pick a small open disc centered at $f(\lambda)$, say U, so that one can find an analytic function g on U satisfying

$$\lambda\epsilon g(U)\subset G,$$

and

$$g(f(w))=w$$

for all $w\epsilon g(U)$. The circle centered at the origin with radius ε divides U into two regions U_- and U_+ where

$$U_- = \{|\beta|<\varepsilon\}\cap U$$

and

$$U_+ = \{|\beta|>\varepsilon\}\cap U.$$

Clearly, $g(U_-)\subset\mathbb{C}\setminus F$ and $\lambda\in\partial g(U_-)$. Hence, the claim is established.

Now, using the claim, we see that if $\lambda\in\partial F\cap G$ and $f'(\lambda)\neq 0$, then, for sufficiently small δ, the set int $\Delta(\lambda,\delta)\setminus F$ contains an arc whose diameter exceeds $\delta/2$. Hence, for such a λ and all sufficiently small δ,

$$\gamma(\Delta(\lambda,\delta)\setminus F) \geq \delta/8$$

because the analytic capacity of a continuum is bigger than or equal to 1/4 its diameter [19, p.199]. Also recall the analytic capacity of a disc of radius δ is δ [19, p. 196]. Hence if $\lambda\in\partial F\cap G$ and $f'(\lambda)\neq 0$, then

$$\liminf_{\delta\to 0} \frac{\gamma(\Delta(\lambda,\delta)\setminus F)}{\gamma(\Delta(\lambda,\delta)\cap\partial H)} \geq 1/8$$

(We have used the obvious fact that γ is monotone.)

Let E be a σ-curvilinear null set satisfying condition (iii) of our hypothesis. We then have, for each $z\in\partial K\setminus E$, there exists $r\geq 1$ so that

$$\liminf_{\delta\to 0} \frac{\gamma(\Delta(z,r\delta)\setminus F)}{\gamma(\Delta(z,\delta)\cap\partial H)} \geq \liminf_{\delta\to 0} \frac{\gamma(\Delta(z,r\delta)\setminus K)}{\delta}>0.$$

(Recall $F\subset K$.)

Let $\widetilde{E}$ be the union of E with the set $\{\lambda\in\partial F\cap G\colon f'(\lambda)=0\}$. Clearly, $\widetilde{E}$ is a σ-curvilinear null set, and it satisfies condition (vi) of Theorem 25 for the compact set F. Hence, we see R(F) is strongly pointwise boundedly dense in $H^\infty(H)$ if we apply Davie's result [14] again.

Let $\{r_n\}$ be a sequence of rational functions in R(F) with $\|r_n\|\leq\|1/f\|$ (sup norms computed on H) and $r_n\to 1/f$ pointwise everywhere on H. Passing to a subsequence if need be, we assume that there exists $h\in R^\infty(F,\mu)$ such that $r_n\to h$ weak-star in $L^\infty(\mu)$. We now show that

$$k \equiv \Lambda(f)\, h - 1$$

is zero almost everywhere μ. (For the rest of the proof any nomenclature used, that has not been defined here, can be found in [11 or 10].)

Since S is pure, it readily follows that $R^{\infty}(F,\mu)$ has no L^{∞} direct summand. Hence, there is a complex measure $\nu \in L^1(\mu) \cap R(F)^{\perp}$ such that ν and μ are mutually absolutely continuous, $\nu \equiv \mu$.

Let C_ν denote the set of points where the Newtonian potential of ν exists. That is, $w \in C_\nu$ if

$$\int_F \frac{d|\nu|(z)}{|z-w|} < \infty.$$

If $w \in C_\nu$, then the Cauchy transform of ν at w is defined by

$$\hat{\nu}(w) \equiv \int_F \frac{d\nu(z)}{z-w}.$$

The Cauchy transform of ν at w is defined planar a.e. ([19, p. 46]).

Repeating verbatim the argument given in [11] on the bottom of page 41, we see that $(k\nu)\hat{}(w)=0$ whenever $w \in C_\nu$ and $\hat{\nu}(w)=0$.

Using [23, Theorem 2 or 24, Theorem 4], we see $R(F)=A(F)$. Hence, the set of points on ∂F which are not peak points for R(F) has planar measure zero (cf. [39, Chapter 3, Section 6, Theorem 2 and the remark preceding it]). If $w \in \partial F \cap C_\nu$ and w is a peak point for R(F), then $\hat{\nu}(w)=0$; otherwise, by using [19, Theorem 11.3, p. 54] and noting that

$$\frac{1}{\hat{\nu}(w)} \frac{d\nu}{(\chi-w)}$$

is a complex representing measure for R(F) at w, we see $\nu(\{w\})\neq 0$. This contradicts the fact that $w \in C_\nu$, so we can conclude $(k\nu)\hat{}(w)=0$ almost everywhere (planar measure) on ∂F.

Now suppose $w \in H \cap C_\nu$ and $\hat{\nu}(w)\neq 0$. Then the map $\phi \to \hat{\phi}(w)$, where

$$\hat{\phi}(w) = \frac{1}{\hat{\nu}(w)} \int \frac{\phi(z)\, d\nu(z)}{z-w},$$

is a weak-star continuous, multiplicative linear functional on $R^{\infty}(F,\mu)$ such that $\hat{r}(w)=r(w)$ for each $r\in R(F)$. Choose a sequence $\{s_n\}$ in R(K) such $\|s_n\|\le\|f\|$, and $s_n\to f$ pointwise on G. Noting that $\Lambda(s_n)=s_n$, we see that $s_n\to\Lambda(f)$ weak-star, and, therefore,

$$s_n(w) = \hat{s}_n(w)\to(\Lambda(f))^{\wedge}(w)$$

for each $w\in H\cap C_{\nu}$ satisfying $\hat{\nu}(w)\neq 0$. Hence, for each of these w's we see $(\Lambda(f))^{\wedge}(w)=f(w)$. Now,

$$\hat{h}(w) = (f^{-1})(w)$$

because the sequence $\{r_n\}$ converges to f^{-1} strongly pointwise boundedly on H. Since $f(w)r_n(w)\to 1$, it follows that, for each $w\in H\cap C_{\nu}$ satisfying $\hat{\nu}(w)\neq 0$,

$$\hat{k}(w) = \hat{f}(w)\hat{h}(w)-1 = 0.$$

We have now shown that $(k\nu)^{\wedge}(w)=0$ on $(C_{\nu}\cap H)\cup(C_{\nu}\cap$ peak points$)$ $\cup(C\setminus F)$. (One should recall $\nu\perp R(F)$; whence, $\hat{\nu}(w)=0$ for all $w\in C\setminus F$.) Observing that the complement of this last set has planar measure zero and then using [19, Corollary 8.3, p. 47], we see that $k\nu=0$. Therefore, $k=0$ almost everywhere μ (as we promised to verify).

It now follows from the definition of k that

$$0 = \rho(\Lambda(f))\rho(h)-1.$$

Hence,

$$1 = \pi(f)\rho(h);$$

therefore, $\pi(f)$ is invertible and the proof is completed.∎

<u>Remark</u> 26. Suppose S is a pure subnormal operator and K is a compact set containing $\sigma(S)$. If $G\equiv \operatorname{int} K$ and conditions (ii) and (iii) hold in Theorem 24, then the proof of that theorem exhibits a unital representation (that is unique via Theorem 10) π defined on $H^{\infty}(G)$.

<u>Remark</u> 27. Suppose π is a unital representation defined on $H^{\infty}(G)$ such that $S=\pi(\chi)$ is pure and

(28) $\sigma(S) \cap G$ is a dominating set for $H^\infty(G)$.

If Conjecture 12 is true, then, under this additional hypothesis (28), π is an isometry; i.e., $\|\pi(f)\| = \|f\|$ for every $f \in H^\infty(G)$. (This last implication is established in the proof of Theorem 24.) We have not even been able to show that each π satisfying the properties in the first sentence of this remark has this isometric property.

Remark 29. The fundamental strategy of the proof of Theorem 24 is very simple: For any nonconstant $f \in H^\infty(G)$, if $H = \{z \in G: |f(z)| > \varepsilon\}$, then extend π to a representation $\tilde{\pi}: H^\infty(H) \to \mathcal{B}(\mathcal{H})$. (That is, $\tilde{\pi}(g) = \pi(g)$ for all $g \in H^\infty(G)$). The complications enter in the construction of $\tilde{\pi}$. There are several extension problems lurking in the background. For example,

Question 30. Suppose π is a unital representation defined on $H^\infty(G)$ with $S = \pi(\chi)$ pure. Does there exist an extension $\tilde{\pi}$ of π defined on $H^\infty(H)$, where H is an open subset of G such that $H \cap \sigma(S)$ is a dominating subset for $H^\infty(H)$?

We now turn to the other natural problem of relating the values of f on $G \cap \sigma_e(S)$ to the set $\sigma_e(\pi(f))$. Recall that if $T \in \mathcal{B}(\mathcal{H})$, then $\sigma_e(T)$ denotes the essential spectrum of T.

Throughout the rest of this section we shall assume π is a (fixed) unital representation defined on $H^\infty(G)$ such that $G \cap \sigma(S)$ is a dominating set for $H^\infty(G)$ where, of course, $S = \pi(\chi)$.

Lemma 32. $\partial G \subset \sigma_e(S)$.

Actually, we prove that ∂G is contained in the set $\sigma_{\ell e}(S)$, the left essential spectrum of S. For any operator T, the set of $\lambda \in \mathbb{C}$ for which there exists an orthonormal sequence $\{x_n\}$ of unit vectors such that $\|(T-\lambda)x_n\| \to 0$ is the left essential spectrum of T. (Consult [34, Proposition 2.15].)

Proof of Lemma 32. Using a unitary operator, if necessary, we may assume, as done in Lemma 5, that N, the minimal normal extension of S, is $\underset{i}{\oplus} M_\chi$ on the space $\underset{i}{\oplus} L^2(\mu_i)$, where

$\mu_{i+1} << \mu_i$ for all i. Let E be the spectral measure for N.

Suppose $0 \in \partial G$. By our standing assumption 0 is an essential boundary point for $H^\infty(G)$. Fix a small $\delta > 0$. By Propositions 5.2 and 5.3 in [18,pp. 64–65] we see that 0 is an essential boundary point for $H^\infty(G \cup \{|z| > \delta\})$. Therefore, by Corollary 7.2 in [18, p. 73], there exists $f \in H^\infty(G \cup \{|z| > \delta\})$ such that

$$\limsup \{|f(z)| : z \to 0, z \in G\} = 1,$$

and, for each $\lambda \in \partial G$ with $0 \neq \lambda$,

$$\limsup \{|f(z)| : z \to \lambda, z \in G\} < 1.$$

Raising f to a sufficiently high power, we may assume there exists $f \in H^\infty(G \cup \{|z| > \delta\})$ such that $\|f\| = 1$ and $|f(z)| < 1/4$ on the set $\{|z| > 2\delta\}$. Since $G \cap \sigma(S)$ is a dominating set for $H^\infty(G)$, there exists an infinite sequence $\{\lambda_n\}$ in $\sigma(S) \cap \{|z| < \delta\}$ such that

$$|f(\lambda_n)| \to 1.$$

By Lemma 21, we see $f(\lambda_n) \in \sigma(\pi(f))$ for each n. It follows then there is some point $\tau \in \partial \sigma(\pi(f))$ with $|\tau| = 1$ such that τ is not an isolated point of $\sigma(\pi(f))$. Therefore $\tau \in \sigma_{\ell e}(\pi(f))$ by the Putnam–Schecter theorem [17, consult the footnote to Theorem 3.3]. Let $\{x_n\}$ be an orthonormal sequence of unit vectors in $\mathcal{H}$ (the space on which S acts) such that

(33) $\|(\pi(f) - \tau)x_n\| \to 0.$

Now f is analytic on $\{|z| > \delta\}$ and $|f - \tau| \geq 1/2$ on $\{|z| > 2\delta\}$. Combining these two facts with (33) and Lemma 5, one readily sees that

$$\|E(\{|z| > 2\delta\})x_n\| \to 0.$$

Thus, there exists an integer N such that, for all $n \geq N$,

$$\|Sx_n\| \leq 3\delta.$$

The result now follows from Condition 8 in Theorem 1.1 of [17].∎

The organization of our results, relating the values of f on $G \cap \sigma(S)$ and the set $\sigma_e(\pi(f))$, follows very closely that used in the first part of this section. We first present a general inclusion (analogous to Lemma 21).

Lemma 34. For each $f \in H^\infty(G)$, we have

$\sigma_e(\pi(f)) \supset \{w \in \mathbb{C}$: there exists a sequence $\{\lambda_n\} \subset G \cap \sigma(S)$ with $\lambda_n \to \lambda$, where $\lambda \in \sigma_e(S)$, and $f(\lambda_n) \to w\}$.

The proof of Lemma 34 hinges on a theorem of Gamelin and Garnett [22]. (This result also appears in [18, Theorem 4.5, p.187]). The idea that that theorem is applicable to this problem is due to Axler [3]. In addition, we should mention that some of arguments in the various cases presented in the proof are taken directly from the last paper.

For completeness, we state the result of Gamelin and Garnett referred to in the last paragraph. Let $\mathcal{M}(G)$ denote the maximal ideal space of $H^\infty(G)$. The fiber $\mathcal{M}_\lambda(G)$ above a point $\lambda \in G$ consists of all those $\phi \in \mathcal{M}(G)$ such that $\phi(\chi)=\lambda$. If $f \in H^\infty(G)$ extends analytically to a neighborhood of λ, then $\phi(f)=f(\lambda)$. If $\{z_n\}$ is a sequence of points in G, we say that $\{z_n\}$ is an interpolating sequence for $H^\infty(G)$ if, for every bounded sequence $\{s_n\} \in \ell^\infty$, there exists $f \in H^\infty(G)$ such that $f(z_n)=s_n$.

Theorem 35 (Gamelin–Garnett). Suppose $\{z_n\}$ is a sequence in G which converges to a point $\lambda \in \partial G$. Then either $\{z_n\}$ has a subsequence which is an interpolating sequence, or else the sequence $\{f(z_n)\}$ converges for all $f \in H^\infty(G)$. In the latter case, there exists a unique $\phi_\lambda = \phi \in \mathcal{M}_\lambda(G)$, called the distinguished homomorphism in the fiber $\mathcal{M}_\lambda(G)$, such that

$$\phi_\lambda(f) = \lim_{n\to\infty} f(z_n)$$

for all $f \in H^\infty(G)$. (Uniqueness means ϕ_λ is independent of the sequence $\{z_n\}$; if $\{w_n\}$ is another sequence in G that converges to λ and $\lim_{n\to\infty} f(w_n)$ exists for all $f \in H^\infty(G)$, then $\lim f(z_n) = \lim f(w_n)$ for every $f \in H^\infty(G)$.) Furthermore, if the fiber $\mathcal{M}_\lambda(G)$ has a distinguished homomorphism, then $\{\lambda\}$ is a component of $\mathbb{C} \setminus G$.

Proof of Lemma 34. Fix a sequence $\{z_n\}$ in $G\cap\sigma(S)$ and a nonconstant function f in $H^\infty(G)$ such that $z_n\to z_0\in\sigma_e(S)$ and $f(z_n)\to 0$. We have to show $0\in\sigma_e(\pi(f))$; that is, $\pi(f)$ is not a Fredholm operator. Suppose to the contrary, $\pi(f)$ is invertible in the Calkin algebra. Then there exists an $\epsilon_0>0$ such that $\pi(f)-\beta$ is Fredholm for all $|\beta|<\epsilon_0$ (because the invertible elements of Banach algebra are open).

If $z_0\in G$, then we can find a function $g\in H^\infty(G)$ such that

$$f = (\chi-z_0)g.$$

Hence,

$$\pi(f) = (S-z_0)\pi(g),$$

which clearly implies $\pi(f)$ is not Fredholm, a contradiction to our supposition. Therefore, we may (and do) assume $z_0\in\partial G$. Using this same argument, we see that we may also assume that $z_n\notin\sigma_e(S)$ for all n.

Now let us consider the case that $\{z_n\}$ contains a subsequence that is interpolating. Relabeling, if need be, we assume the subsequence is the sequence $\{z_n\}$. The map $g\to\{g(z_n)\}$ is a bounded linear map of $H^\infty(G)$ onto ℓ^∞. Hence, by the open mapping theorem, for each bounded sequence $\{\alpha_n\}$ there exists a function $g\in H^\infty(G)$ such that

$$\|g\| \le C\|\{\alpha_n\}\|$$

with $g(z_n)=\alpha_n$ for all n, where C is a (universal) constant.

For each integer N, there is a function $f_N\in H^\infty(G)$ such that

$$f_N(z_n) = \begin{cases} 0 & \text{if } n < N \\ f(z_n) & \text{if } n \ge N \end{cases}$$

and $\|f_N\|\le C\ \sup\{|f(z_n)|:n\ge N\}$. Hence $\|f_N\|\to 0$; since π is continuous, we then see $\pi(f-f_N)\to\pi(f)$ in the norm topology of $\mathcal{B}(\mathcal{H})$. Using the argument in the second paragraph of this proof, we can find a sequence $\{g_{nN}\}_{n=N}^{\infty}\subset H^\infty(G)$ such that

$$\pi(f-f_N) = (S-z_n)\pi(g_{nN})$$

for all $n \geq N$. Therefore, the kernel of $\pi(f-f_N)^*$, denoted $\ker(\pi(f-f_N))^*$, contains the closed linear span of the elements in $\ker(S-z_n)^*$ for all $n \geq N$. Recalling that, $\{z_n\} \subset \sigma(S) \cap G$ is an infinite sequence ($z_n \to z_0 \in \partial G$), and $\|Sx\| \geq \|S^*x\|$ for all $x \in \mathcal{H}$ (subnormal operators are hyponormal), we easily see that $\ker(\pi(f-f_N))^*$ is infinite dimensional. Hence, $\pi(f-f_N)$ is not Fredholm; whence, $\pi(f)$ is not Fredholm, a contradiction to our supposition.

Therefore, by Theorem 35, it must be the case that there exists a distinguished homomorphism $\phi_{z_0} \in \mathcal{M}_{z_0}(G)$; so

$$g(z_n) \to \phi_{z_0}(g)$$

for all $g \in H^\infty(G)$. We divide this case into two parts.

First, let us suppose $f|_{\sigma(s)\cap G}$ is continuous at z_0; i.e., for $z \in \sigma(S) \cap G$,

$$\lim_{z \to z_0} f(z) = 0.$$

We claim that for every $\lambda \in \partial G$ and every $\varepsilon > 0$, there exists a $\beta \in \partial G$ such that $|\beta - \lambda| < \epsilon$ and there exists a sequence $\{w_n\} \subset G \cap \sigma(S)$ such that $w_n \to \beta$ and $\{w_n\}$ is an interpolating sequence for $H^\infty(G)$. Postponing the proof of the claim until the next paragraph, we finish the argument of the case at hand. Choose $\delta > 0$ such that $|f(z)| < \varepsilon_0/2$, whenever $|z - z_0| < \delta$ and $z \in G \cap \sigma(S)$. (Recall ϵ_0 was chosen in the first paragraph of the proof.) Choose $\beta \in \partial G$ and a sequence $\{w_n\}$ in $G \cap \sigma(S)$ such that $|\beta - z_0| < \delta/2$, $w_n \to \beta$, and $\{w_n\}$ is an interpolating sequence for $H^\infty(G)$. From Lemma 32, we see $\beta \in \sigma_e(S)$. Moreover, if τ is any cluster point of the sequence $\{f(w_n)\}$, then $|\tau| \leq (\varepsilon_0)/2$. But, a repetition of the argument in first two cases of this proof shows $\tau \in \sigma_e(\pi(f))$, contradicting the choice of ε_0.

We now prove the claim in the last paragraph. Fix $\lambda \in \partial G$ and $\varepsilon > 0$. From the proof of Lemma 32, we see there exists a function $g \in H^\infty(G)$ such that $\|g\|=1$ and $|g(w)|<\varepsilon$ for all $w \in (C \setminus \Delta(\lambda,\varepsilon/2)) \cap G$. Since $G \cap \sigma(S)$ is dominating, there exists an infinite sequence $\{w_n\} \subset G \cap \sigma(S) \cap \Delta(\lambda,\epsilon/2)$ such that $|g(w_n)| \to 1$. Without loss of generality, we may assume that $w_n \to \beta \in \partial G$. Clearly $|\lambda-\beta|<\varepsilon$. We now show that some subsequence of $\{w_n\}$ is interpolating.

Suppose to the contrary, there exists a distinguished homomorphism $\phi_\beta \in M_\beta(G)$ and

$$\|g\| = 1 = |\phi_\beta(g)|.$$

Using another result in [22], we see there exists a positive measure ν_β with compact support in the plane, $\|\nu_\beta\|=1$, with the property that

$$\phi_\beta(h) = \int_G h\,d\nu_\beta$$

for all $h \in H^\infty(G)$. (Note, ν_β is carried by G.) Observing that $\|g\|-|g(w)|>0$ for all $w \in G$, we see

$$1 = \|g\| = |\phi_\beta(g)| < \int_G \|g\|\,d\nu_\beta = 1;$$

this absurdity shows our initial assumption (in this paragraph) is false. The claim is established.

We are left with the case that $g(z_n) \to \phi_{z_0}(g)$ for all $g \in H^\infty(G)$ and $f|_{G\cap\sigma(s)}$ is not continuous at z_0. Therefore, we can choose a sequence $\{w_n\} \subset G \cap \sigma(S)$ such that $w_n \to z_0$ and $f(w_n) \to \tau$ where $\tau \neq 0$. It must be the case then that $|\tau| \geq \varepsilon_0$. (Reasons, $\{w_n\}$ contains a subsequence that is interpolating for $H^\infty(G)$ because $\phi_{z_0}(f)=0$. Whence, by the argument in the first two cases, $\tau \in \sigma_e(\pi(f))$. The statement now follows from the choice of ε_0.)

Fix $\varepsilon>0$ where $\varepsilon<\varepsilon_0$. There exists a neighborhood V_ε of z_0 such that

$\phi = V_{\varepsilon}\cap G\cap\sigma(S)\cap\{z\in G: \varepsilon/4\leq|f(z)|\leq\varepsilon/2\}.$

(To see this, argue by contradiction and use tha argument in the preceding paragraph.) Using [3, Lemma 22], we may choose a subset V_{ε}' of V_{ε} containing z_0 in its interior such that $V_{\varepsilon}'\cap G$ is pathwise connected. For n sufficiently large, z_n and w_n belong to V_{ε}' and

$$|f(z_n)| < \varepsilon/4,$$

$$|f(w_n)| \geq \varepsilon.$$

Fix a large n. Let γ_{ε} be a path joining z_n and w_n that lies in $V_{\varepsilon}'\cap G$, parameterized in such a way that $\gamma_{\varepsilon}(0)=z_n$ and $\gamma_{\varepsilon}(1)=w_n$.

Define t_0 as follows:

$$t_0 \equiv \inf\{t\in[0,1]:\gamma_{\varepsilon}(t)\notin\sigma(S)\}.$$

(The set on which we compute the infimum is nonempty, because $\gamma_{\varepsilon}(t)\notin\sigma(S)$ when

$$\varepsilon/4 \leq |f(\gamma_{\varepsilon}(t))| \leq \varepsilon/2.)$$

Clearly, $|f(\gamma_{\varepsilon}(t_0))|<\varepsilon/4$ and $\gamma_{\epsilon}(t_0)\in\partial\sigma(S)$. If $\gamma_{\epsilon}(t_0)\in\sigma_e(S)$, then, repeating the argument in the second paragraph of the proof, we see $f(\gamma_{\varepsilon}(t_0))\in\sigma_e(\pi(f))$. This is a contradiction to the fact $\pi(f)-\beta$ is Fredholm for all $|\beta|<\varepsilon_0$. It must be then the case that $\gamma_{\varepsilon}(t_0)$ is an isolated eigenvalue of $\sigma(S)$ with finite multiplicity ([17]). (If we had assumed S was a pure subnormal operator, then this last case cannot happen; eigenspaces reduce subnormal operators and their restrictions to such subspaces are normal. The proof of Lemma 34 would be completed.)

Observing that the kernel of $\pi(f)-f(\gamma_{\varepsilon}(t_0))$ contains the kernel of $S-\gamma_{\varepsilon}(t_0)$, we see $f(\gamma_{\epsilon}(t_0))$ is an eigenvalue of finite multiplicity of $\pi(f)$. (Finite because $\gamma_{\varepsilon}(t_0)\notin\sigma_e(\pi(f))$.) Now letting $\varepsilon\to 0$, we can construct a sequence $\{f(\gamma_{\varepsilon_n}(t_0))\}_{n=1}^{\infty}$ of distinct eigenvalues of $\pi(f)$ such that $f(\gamma_{\varepsilon_n}(t_0))\to 0$ as $n\to\infty$.

Therefore, from [17], $0 \in \sigma_e(\pi(f))$; a contradiction to our supposition $\pi(f)$ is Fredholm. There are no other cases left to consider; Lemma 34 has been established.∎

As mentioned earlier, Lemma 34 is the Calkin algebra analogue of Lemma 21. Before we give the analogue of Corollary 22, we introduce some notation that will be useful to us here and in the next two sections. If $f \in H^\infty(G)$ and $\lambda \in \partial G$, then the cluster set of f at λ, written $C(f,\lambda)$, consists of all points $w \in \mathbb{C}$ for which there exists a sequence $\{\lambda_n\} \subset G$ satisfying $\lambda_n \to \lambda$ and $f(\lambda_n) \to w$. Recall that $\mathcal{M}(G)$ denotes the maximal ideal space of $H^\infty(G)$. The image of f under the Gelfand map is denoted by $\hat{f}$; that is,

$$\hat{f}(\phi) = \phi(f)$$

for all $\phi \in \mathcal{M}(G)$. Consulting [18, Theorem 3.4, p. 176], we see

$$\hat{f}(\mathcal{M}_\lambda) = C(f,\lambda),$$

where, as before, $\mathcal{M}_\lambda = \mathcal{M}_\lambda(G)$ is the fiber of $\mathcal{M}(G)$ over λ. Let

$$\mathcal{M} \equiv \bigcup_{\lambda \in \partial G} \mathcal{M}_\lambda .$$

Corollary 36. Let π be a unital representation defined on $H^\infty(G)$ with $\sigma(S) = \bar{G}$ (where, as always, $S = \pi(\chi)$). If $f \in H^\infty(G)$, then

$$\sigma_e(\pi(f)) = \hat{f}(\mathcal{M}) \cup f(G \cap \sigma_e(S)).$$

Proof. Let B_f denote the set on the right-hand side of the equality we want to verify. Clearly $\sigma(S) \cap G$ is a dominating set for $H^\infty(G)$; whence,

$$\sigma_e(\pi(f)) \supset B_f$$

from Lemma 34.

So suppose $0 \notin B_f$, we have to show $\pi(f)$ is Fredholm. Without loss of generality, we may assume $0 \in f(G)^-$. Since $0 \notin B_f$, it follows that f has only a finite number of zeros $\{z_1, z_2, \ldots, z_n\} \subset G$; furthermore, $\{z_1, \ldots, z_n\} \subset G \setminus \sigma_e(S)$. We assume that we have listed all the zeros of f according to their multiplicities. Hence, there exists a function $g \in H^\infty(G)$ such that

$$f = (\pi_i (\chi-\lambda_i))g$$

and $g(z)\neq 0$ for all $z\in G$. Using the fact that $0\notin B_f$ again, we see $g^{-1}\in H^\infty(G)$. Since

$$\pi(f) = \pi(g) \pi_i (S-\lambda_i),$$

one sees that $\pi(f)$ is Fredholm. ∎

Perhaps it is worth noting that in our proof above that shows $\sigma_e(\pi(f))\supset B_f$, the subnormality of S is not used. It is clear that a majority of these results can be generalized to a larger class of operators. How large of a class is not apparent to us. We now present the Calkin algebra analogue of Theorem 24.

Theorem 37. Let π be a unital representation defined on $H^\infty(G)$, satisfying the hypotheses of Theorem 24. With these assumptions we have, for each $f\in H^\infty(G)$,

$\sigma_e(\pi(f)) = \{w\in\mathbb{C}$: there exist $\lambda\in\sigma_e(S)$ and

a sequence $\{\lambda_n\}\subset G\cap\sigma(S)$ satisfying $\lambda_n\to\lambda$ and $f(\lambda_n)\to w\}$.

Proof. Let R_f denote the set on the right-hand side of the equality we want to verify. From our hypothesis, Lemma 34 is applicable. Therefore, we only need to show $R_f\supset\sigma_e(\pi(f))$.

So suppose $0\notin R_f$, we need to verify that $\pi(f)$ is Fredholm. Looking at Theorem 24, we see that we might as well assume $0\in f(G\cap\sigma(S))^-$ too. From Lemma 32, we see we can choose a positive integer m such that for each integer i, $1\leq i\leq m$, there is a zero of f, say z_i, belonging to $(G\cap\sigma(S))\setminus\sigma_e(S)$, and a positive number r_i such that $\Delta(z_i,r_i)^-\subset G$, satisfying $|f|\geq\delta$ on the set

$$(G\cap\sigma(S))\setminus\bigcup_{i=1}^{m}\Delta(z_i,r_i)$$

where δ is a small positive number. Note the set just displayed is dominating for $H^\infty(G)$.

Let ε_i, for $1\leq i\leq m$, be defined by

$$\varepsilon_i = \sup\{|f(z)|: z \in \Delta(z_i, r_i)\},$$

and set $\varepsilon = \min_{1 \leq 1 \leq m} (\varepsilon_i)/2$. Clearly, $0 < \varepsilon < \delta$. Let $H = \{z \in G: |f(z)| > \varepsilon\}$ and $F = H^-$. We note that $\partial\Delta(z_i, r_i) \subset \mathrm{int}' F = H$. As was done in the proof of Theorem 24, one can show that F satisfies condition (vi) of Theorem 25. Let

$$\widetilde{F} \equiv F \cup \left(\bigcup_{i=1}^{m} \cup \Delta(z_i, r_i)\right).$$

One easily checks that

$$\widetilde{H} \equiv \mathrm{int}\, \widetilde{F} = H \cup \left(\bigcup_{i=1}^{m} \Delta(z_i, r_i)\right).$$

It is also easy to see that for each $z \in \partial\widetilde{F}$ and for all sufficiently small β

$$\Delta(z, \beta) \setminus F = \Delta(z, \beta) \setminus \widetilde{F},$$

and

$$\Delta(z, \beta) \cap \partial H = \Delta(z, \beta) \cap \partial\widetilde{H}.$$

With these facts, its apparent then that $\widetilde{F}$ satisfies condition (vi) of Theorem 25.

Using the same technique as done in Theorem 24, we can construct a unital representation $\widetilde{\pi}$ from $H^\infty(\widetilde{H})$ to $\mathcal{B}(\mathcal{H})$ such that $\widetilde{\pi}(g) = \pi(g)$ for all $g \in H^\infty(G)$. From the definition of the open set $\widetilde{H}$, clearly we can find integers p_i, for $1 \leq 1 \leq m$, and an invertible function $g \in H^\infty(\widetilde{H})$ such that

$$f(w) = g(w) \prod_{i=1}^{m} (w - z_i)^{p_i}$$

for all $w \in \widetilde{H}$. Hence,

$$\pi(f) = \widetilde{\pi}(g) \prod_{i=1}^{m} (S - z_i)^{p_i}.$$

Observing that $\widetilde{\pi}(g)$ is invertible and $z_i \notin \sigma_e(S)$ for $1 \leq i \leq m$, we see that $\pi(f)$ is a Fredholm operator.∎

Remark 38. Before we close this section, one may wonder how crucial to our results was the hypothesis that π, a unital representation, was defined on $H^\infty(G)$ rather than some other Banach algebra of functions. One important property of $H^\infty(G)$, that was used explicitly and implicitly many times, is that $H^\infty(G)$ is invariant under localization. That is to say, if ϕ is any smooth function (continuous partial derivatives) defined on $\mathbb{C}$ with compact support and if $f \in H^\infty(G)$, then the function $T_\phi(f) \in H^\infty(G)$ where

$$T_\phi(f)(w) = 1/\pi \iint \frac{f(w)-f(z)}{w-z} \frac{\partial \phi}{\partial \bar{z}} \, dxdy$$

for all $w \in \mathbb{C}$.

Following [21, or 9, Section 17], we say that a closed subalgebra A of C(K) is T-invariant if the following two conditions are met (K is a compact subset of $\mathbb{C}$):

(xi) $R(K) \subset A$,

(xii) $T_\phi A \subset A$ for all smooth ϕ with compact support.

Now suppose S is a pure subnormal operator and A is a T-invariant subalgebra of C(K) where $K \supset \sigma(S)$. Assume further, π is a unital representation defined on A with $\pi(\chi)=S$. Using the methods of Section 2 and some elementary facts about T-invariant algebras, we can show that

$$\pi(f) = f(N)|_{\mathcal{H}}$$

for all $f \in A$. (Here N is the minimal normal extension of S and the latter operator is defined on $\mathcal{K}$. The operator f(N) is defined because f is continuous on $\sigma(N) \subset \sigma(S) \subset K$. One has to show, in addition to the indicated equality, that $f(N)\mathcal{H} \subset \mathcal{H}$.) Now let Q denote the set of nonpeak points of A, and λ_Q the area measure on Q. Without giving the details, we can extend π to a unital representation defined on $H^\infty(\lambda_Q)$, the weak-star closure of A in $L^\infty(\lambda_Q)$. The algebra of functions $H^\infty(\lambda_Q)$ is invariant under localization. It is natural to inquire about the spectral mapping properties for the functional calculus that this extension of π induces. We plan on pursuing this line of thought in a future paper.

CHAPTER V

REPRESENTATIONS OF

$H^\infty(G)$ INTO $L^\infty(\mu)$

Throughout this section G will be a bounded region in the plane $\mathbb{C}$ and μ will be a probability measure with spt $\mu \subset \bar{G}$. In this section we characterize those unital representations $\pi: H^\infty(G) \to L^\infty(\mu)$ such that $\pi(\chi) = \chi$ where, as in the preceding sections, χ is the function defined by $\chi(z) = z$ for all z. (Recall that, by definition, a unital representation π is a continuous algebra homomorphism with the property that $\pi(1) = 1$. Let us note that, with our hypothesis that $\pi(\chi) = \chi$, the assumption of continuity is redundant; consult the remarks at the end of Section 1. Therefore, this section characterizes all algebra homomorphisms π of $H^\infty(G)$ into $L^\infty(\mu)$ satisfying $\pi(1) = 1$ and $\pi(\chi) = \chi$.) Our characterization will describe a one-to-one correspondence between these representations and certain measures on $\mathcal{M}(G)$, the maximal ideal space of $H^\infty(G)$. If π is such a unital representation, then for all $f \in H^\infty(G)$

$$\pi(f)(z) = f(z)$$

$\mu|_{\underline{G}}$ almost everywhere (consult Lemma 5). Thus <u>we shall henceforth assume in this section</u> that $\mu(G) = 0$. Using our characterization, we shall obtain the following example that answers a question in [8].

<u>Example</u> 40. Let G=D and suppose spt $\mu \subset \partial D$. Let B be a Blaschke product whose zeros accumulate at every point of ∂D. Then for each $f \in L^\infty(\mu)$ with $\|f\| \leq 1$, there exists a unital representation $\pi: H^\infty(G) \to L^\infty(\mu)$ such that $\pi(\chi) = \chi$ and $\pi(B) = f$.

This shows, in particular, that without the hypothesis of purity on S, many unital representations exist that map χ to S.

Example 40 can be used to show that Corollary 11 (in Section II) is close to the best possible result. We now construct a unital representation π: $H^\infty(D) \to \mathcal{B}(\mathcal{H})$ such that $\pi(\chi) = W$ where W is the bilateral shift of multiplicity two and ran $\pi \not\supset \{W\}''$. Recall that m is normalized Lebesgue measure on ∂D.

Example 41. Let W be the bilateral shift of multiplicity two; i.e., $W = M_z \oplus M_z$ on $L^2(m) \oplus L^2(m)$. Choose two functions $f_i \in L^\infty(m)$ such that $\|f_i\| \leq 1$ and $f_1 \neq f_2$. Let B be an infinite Blaschke product whose zeros accumulate at every point on ∂D. From Example 40 there exist two unital representations π_i: $H^\infty(D) \to \mathcal{B}(L^2(m))$ such that $\pi_i(\chi) = M_z$ on $L^2(m)$ and $\pi_i(B) = M_{f_i}$ on $L^2(m)$.

Define π: $H^\infty(D) \to \mathcal{B}(L^2(m) \oplus L^2(m))$

via

$$\pi(g) = \begin{bmatrix} \pi_1(g) & 0 \\ 0 & \pi_2(g) \end{bmatrix}$$

for all $g \in H^\infty(D)$. Since $\mathcal{W}^*(W) = \{W\}'' = \{M_f \oplus M_f\colon f \in L^\infty(m)\}$, clearly $\pi(B) \notin \{W\}''$.∎

For the case of the open unit disc and normalized Lebesgue measure, we shall go further in our analysis of representations. We shall characterize these representations in terms of measurable cross sections of $\mathcal{M}(D)$. For an arbitrary region G and for an arbitrary point $z \in G$, the fiber of $\mathcal{M}(G)$ over z is denoted $\mathcal{M}_z(G)$. Hence,

$$\mathcal{M}_z(G) = \{\phi \in \mathcal{M}(G)\colon \phi(\chi) = z\}.$$

With this notation we can state this later characterization as follows:

Theorem 42. Suppose $\pi: H^\infty(D) \to L^\infty(m)$ is a unital representation such that $\pi(\chi) = \chi$. Then there exists a selector function $s: \partial D \to \mathcal{M}(D)$ such that $s(z) \in \mathcal{M}_z(D)$ for all $z \in \partial D$ and

$$\pi(f)(z) = \hat{f}(s(z))$$

almost everywhere m.

Explanation of the notation $\hat{f}$: if G is a bounded region and $f \in H^\infty(G)$, then $\hat{f}$ is the image of f under the Gelfand transform; i.e., $\hat{f}(\phi) = \phi(f)$ for all $\phi \in \mathcal{M}(G)$. Of course, if a selector function s exists such that $\hat{f}(s(z))$ is a (Lebesgue) measurable function for all

$f \in H^\infty(D)$, then it induces a unital representation that maps χ to χ. Using the axiom of choice, one easily constructs many functions $s:\partial D \to \mathcal{M}(D)$ such that $s(z) \in \mathcal{M}_z(D)$. However, we want to emphasize the fact that in order for such a function to induce a desired representation, one must have $\hat{f}(s(z))$ is a measurable function for all $f \in H^\infty(D)$. (Note, by the first theorem on page 165 of [28], there is no such s such that $\hat{f} \circ s$ is continuous for all $f \in H^\infty(D)$.) The correspondence between representations and measurable cross sections of $\mathcal{M}(D)$ is not one-to-one. We will give an example (consult Example 76) of this in the next section. The authors would like to thank E. Azoff for many conversations in regards to Theorem 42. (As far as the authors can determine this theorem and the preceding example do not follow from the many "selection theorems" that appear in the literature because $\mathcal{M}(D) \setminus D$ is a nonmetrizable, nonseparable compact space.)

<u>We now fix our region G and our probability measure</u> μ with spt $\mu \subset \partial G$. We begin our work by stating an elementary lemma about weak-star convergence of measures. The proof is left to the reader.

<u>Lemma</u> 43. Let X be a compact Hausdorff space and E a closed subset of X. Suppose $\{\nu_\alpha\}$ is a net of measures on X, ν is a measure on X, and $\nu_\alpha \to \nu$ weak-star. If spt $\nu_\alpha \subset E$ for all α, then spt $\nu \subset E$.

For $E \subset \partial G$ let $E^* = \bigcup_{z \in E} \mathcal{M}_z(G)$. Let $\mathcal{M} = (\partial G)^*$. Now assume that E is closed and let $f \in C(E)$. For $z \in E$ and $\phi \in \mathcal{M}_z$ let $\hat{f}(\phi) = f(z)$. Under this identification we may think of $C(E)$ as a closed subspace of $C(E^*)$. Let $p:\mathcal{M} \to \partial G$ be the natural projection: i.e., $p(\phi) = z$ whenever $\phi \in \mathcal{M}_z$. That is, p is the restriction to $\mathcal{M}$ of $\hat{\chi}$, the Gelfand transform of χ.

Let $R = R_\mu$ be the set of positive measures τ on $\mathcal{M}$ such that

$$\int_{\mathcal{M}} \hat{f}\, d\tau = \int_{\partial G} f\, d\mu$$

for all $f \in C(\partial G)$. Note that $R \neq \phi$ by the Hahn-Banach and Riesz representation theorems. The reader can easily check that $\tau \in R$ if and only if $\tau(E^*) = \mu(E)$ for all measurable sets

$E\subset\partial G$. (If E is a measurable set contained in ∂G, then there exists a σ-compact set $F\subset E$ such that $\mu(E\setminus F) = 0$. Consequently, F^* is σ-compact and $\tau(E^*\setminus F^*)=0$; so E^* is measurable.) We define a partial ordering $<$ on R as follows: $\nu<\tau$ if, for all closed subsets E of ∂G, it follows that spt $(\tau\,|\,E^*)\subset$ spt $(\nu\,|\,E^*)$. We define ν and τ to be equivalent if $\nu<\tau$ and $\tau<\nu$. Let $\mathcal{R}$ denote the set of equivalence classes of R. In the obvious way $<$ becomes a partial ordering of $\mathcal{R}$.

Lemma 44. If $\{\nu_\alpha\}$ is a net in R and ν is a measure such that $\nu_\alpha\to\nu$ weak-star, then $\nu_\alpha|_{E^*}\to\nu|_{E^*}$ weak-star for all closed subsets $E\subset\partial G$.

Proof. Let $f\in C(\mathcal{M})$. We must show that

$$\int_{E^*} fd\nu_\alpha\to\int_{E^*} fd\nu.$$

Let $\varepsilon>0$. Choose g in $C(\partial G)$ such that $0\le g\le 1$, $g|_E=1$, and $\int_{\partial G\setminus E} gd\mu < \varepsilon$. By the definition of weak-star convergence there exists an α_0 such that for all $\alpha>\alpha_0$ we have

$$\left|\int_{\mathcal{M}} f\hat{g}d\nu_\alpha - \int_{\mathcal{M}} f\hat{g}d\nu\right| < \varepsilon.$$

Thus, for $\alpha>\alpha_0$

$$\left|\int_{E^*} fd\nu_\alpha - \int_{E^*} fd\nu\right| \le \left|\int_{\mathcal{M}} f\hat{g}d\nu_\alpha - \int_{\mathcal{M}} f\hat{g}d\nu\right|$$

$$+\left|\int_{(\partial G\setminus E)^*} f\hat{g}d\nu_\alpha\right| + \left|\int_{(\partial G\setminus E)^*} f\hat{g}d\nu\right|$$

$$\le \varepsilon + 2\varepsilon\|f\|.$$

Since ε was arbitrary, the proof is completed.■

We now use Zorn's lemma to show that there exist maximal elements (under $<$) in $\mathcal{R}$. Our proof involves the notions of subnets and cluster points of nets and their relationships to compactness; we refer the reader to [40] for the appropriate details.

Theorem 45. There exist maximal elements in $\mathcal{R}$.

Proof. As we have already noted $R\neq\phi$; whence, $\mathcal{R}\neq\phi$. For completeness sake, we spell the details out. Define a bounded linear functional L on $C(\partial G)$ by setting $L(f) = \int fd\mu$ for all f in $C(\partial G)$. Note L has norm one because μ is a probability measure. Identifying $C(\partial G)$ as a closed subspace of $C(\mathcal{M})$, we extend L to be a bounded linear functional of norm one on $C(\mathcal{M})$. By the Riesz representation theorem there exists a measure τ on $\mathcal{M}$ such that $\|\tau\|=1$ and

$$\int_{\mathcal{M}}\hat{f}\,d\tau = \int_{\partial G} f\,d\mu$$

for all $f \in C(\partial G)$. Note that $\tau \geq 0$ because $\|\tau\|=1=\int d\mu=\int d\tau$. Hence $\tau \in R$.

We now apply Zorn's lemma to show that $\mathcal{R}$ has a maximal element. Let ς be a chain in $\mathcal{R}$. For each α in ς let ν_α be a measure in the equivalence class α. Since ς is a directed set with $<$, the set $\{\nu_\alpha\}$ is a net in the space of measures. Noting that $\|\nu_\alpha\|=\int 1\,d\nu_\alpha=\int 1\,d\mu=\|\mu\|=1$, we see that each ν_α lies in the closed unit ball of the space of measures. Since that closed unit ball is weak-star compact, there exists a measure τ and a subnet $\{\tau_\beta\}_{\beta \in B}$ of $\{\nu_\alpha\}$ such that $\tau_\beta \to \tau$ weak-star. From the definitions of $\mathcal{R}$ and weak-star convergence, and by Lemma 44, one easily checks that $\tau \in R$.

We shall show that the equivalence class containing τ is an upper bound for ς. Let $\alpha_0 \in \varsigma$. By the definition of subnet there exists $\beta_0 \in B$ such that $\beta_0 > \alpha_0$. For $\beta > \beta_0$,

$$\operatorname{spt}(\tau_\beta|_{E^*}) \subset \operatorname{spt}(\tau_{\beta_0}|_{E^*})$$

for all closed subsets $E \subset \partial G$. By Lemma 43 and Lemma 44, it now follows that

$$\operatorname{spt}(\tau|_{E^*}) \subset \operatorname{spt}(\nu_{\alpha_0}|_{E^*})$$

for all closed subsets $E \subset \partial G$. Thus, the equivalence class containing τ is an upper bound for ς; whence Zorn's lemma applies and the theorem is proved.∎

Definition 46. We call an element of R a minimal measure if its equivalence class is a maximal element of $\mathcal{R}$.

(Example: if $\mu=\delta_z$, the Dirac measure at a given point $z \in \partial G$, then it is easy to show that

$R = \{\tau: \tau \text{ is a probability measure with spt } \tau \subset \mathcal{M}_z(G)\}$,

and the minimal measures in R consist of the measures δ_φ as φ varies over $\mathcal{M}_z(G)$. Anticipating the discussion in the next paragraph, we point out that each minimal measure $\delta_\varphi \epsilon R$ induces a natural representation of $H^\infty(G)$ into $L^\infty(\mu)$; namely

$$f \to \varphi(f) = \int_{\partial G} \hat{f} d\delta_\varphi.)$$

We now show how a minimal measure ν induces a representation of $H^\infty(G)$ into $L^\infty(\mu)$. Fix $f \epsilon H^\infty(G)$. The mapping $g \to \int \hat{g}\hat{f} d\nu$ defines a bounded linear functional on $C(\partial G)$; thus, there exists a unique complex measure λ on ∂G such that

$$(47) \quad \int_{\mathcal{M}} \hat{g}\hat{f} d\nu = \int_{\partial G} g d\lambda$$

for all g in $C(\partial G)$.

Our goal now is to show that $f \to d\lambda/d\mu$ is the desired representation. To that end, first we shall show that $\lambda << \mu$ and $\|d\lambda/d\mu\|_\infty \le \|f\|$. Let E be a closed subset of ∂G. Let $\{g_n\}$ be a sequence of functions in $C(\partial G)$ such that $0 \le g_n \le 1$, $g_n|_E = 1$, and $g_n \to 0$ pointwise everywhere off E. By the bounded convergence theorem,

$$\lim_n \int g_n d\lambda = \lambda(E).$$

Since

$$\limsup |\int \hat{g}_n \hat{f} d\nu| \le \|f\| \nu(E^*)$$

and $\nu(E^*) = \mu(E)$, it follows from (47) that $|\lambda(E)| \le \|f\| \mu(E)$. Thus, $\lambda << \mu$ and $\|d\lambda/d\mu\|_\infty \le \|f\|$. All that remains is to show that our mapping is multiplicative, and $\chi \to \chi$. The key to these facts (and to the understanding of minimal measures) is the following theorem.

<u>Theorem</u> 48. Let $g = d\lambda/d\mu$ and let E be a closed subset of ∂G such that $g|_E$ is continuous. If $z \epsilon E$ and $\alpha \epsilon \mathcal{M}_z(G) \cap \text{spt}(\nu|_{E^*})$, then $\hat{f}(\alpha) = g(z)$.

The first step in proving Theorem 48 is the following lemma, which will also be used later.

Lemma 49. Let τ be a minimal measure and let ε be a closed subset of $\mathcal{M}$. Define a measure β on ∂G by setting

$$\beta(E) = \tau(E^* \cap \varepsilon)$$

for every measurable set E. Then there exists a measurable subset A of ∂G such that $\beta = \chi_A \mu$. Note that χ_A denotes the characteristic function of the set A (and not the restriction of the function χ to the set A).

Proof. From the definition of β and the definition of minimal measure, one sees $\beta << \mu$ and

$$0 \leq \frac{d\beta}{d\mu} \leq 1.$$

Let $h = d\beta/d\mu$. To prove that h is a characteristic function, it suffices to show if F is a closed subset of ∂G and $h|_F$ (the restriction of h to F) is continuous, then $h^2 = h$ almost everywhere μ on F. Suppose the last assertion is false. Then there exist a closed subset K of F and $\epsilon > 0$ such that $\mu(K) > 0$ and

$$\epsilon \leq h \leq 1 - \epsilon \text{ on } K.$$

Let

$$\sigma \equiv \hat{h}^{-1} \tau|_{K^* \cap \mathcal{E}};$$

that is, for each s in $C(K^*)$

$$\int_{K^* \cap \mathcal{E}} s\, d\sigma = \int_{K^* \cap \mathcal{E}} s\hat{h}^{-1} d\tau.$$

Let

$$\tau' = \sigma + \tau|_{\mathcal{M} \setminus K^*}.$$

By definition $\tau' << \tau$, and $\tau' \epsilon R$ because for every measurable subset H of K,

$$\tau'(H^*) = \sigma(H^*)$$

$$= \int_{H^* \cap \varepsilon} d\sigma$$

$$= \int_{H^* \cap \varepsilon} \hat{h}^{-1} d\tau$$

$$= \int_H h^{-1} d\beta$$

$$= \mu(H).$$

Since τ is a minimal measure, it now follows that τ and τ' must be equivalent. Now $\tau'(K^*)=\tau'(K^* \cap \varepsilon)$ and hence, $\operatorname{spt}(\tau' | _{K^*}) \subset \varepsilon$. But

$$\tau(K^* \cap \varepsilon) = \beta(K)$$

$$\leq (1-\epsilon)\mu(K)$$

$$< \mu(K)$$

$$= \tau(K^*).$$

Hence, $\operatorname{spt}(\tau | _{K^*}) \not\supset \varepsilon$ and we have a contradiction. The lemma has been established.■

<u>Proof of Theorem</u> 48. No harm is done in assuming that $\hat{f}(\alpha)=0$ and $\operatorname{spt}(\mu | _E)=E$. Let $\epsilon>0$. Since $g | _E$ is continuous and $\alpha \in \operatorname{spt}(\nu | _{(E \cap W)^*})$ for all closed neighborhoods W of z, we may assume that $| g(s)-g(t) | < \epsilon$ for all $s,t \in E$. Let $\mathcal{O}=\{\phi \in \mathcal{M}: | \hat{f}(\phi) | < \epsilon\}$. Define a measure β on ∂G by setting

$$\beta F = \nu(F^* \cap (E^* \setminus \mathcal{O}))$$

for every measurable set F. By Lemma 49 there exists a measurable set A such that $\beta=\chi_A \mu$. Since

$$\mathcal{O} \cap \operatorname{spt}(\nu | _{E^*}) \neq \phi,$$

if follows that

$$\nu(E^* \cap \mathcal{O}) > 0$$

and thus,

$$\beta E < \nu E^* = \mu E.$$

Hence, $\chi_A=0$ on some closed subset F of E with $\mu F>0$, and, of course, $\beta F=0$. Clearly

$$\nu(F^*) = \nu(F^* \cap O),$$

so

$$\operatorname{spt}(\nu|_{F^*}) \subset (F^* \cap O)^-$$

and consequently,

$$(50) \quad \operatorname{spt}(\nu|_{F^*}) \subset \{\phi \in M: \ |\hat{f}(\phi)| \le \varepsilon\}.$$

By the definitions of g and λ, we see

$$(51) \quad \int \hat{h}\hat{f} d\nu = \int h d\lambda = \int hg d\mu$$

for all $h \in C(\partial G)$. Let s be an arbitrary function in $C(\partial G)$ and let $\{h_n\}$ be a sequence in $C(\partial G)$ that converges pointwise boundedly to χ_F on ∂G. Replacing h by $h_n s$ in (51) and letting $n \to \infty$, we see that

$$(52) \quad \int_{F^*} \hat{s}\hat{f} d\nu = \int_F s d\lambda = \int_F sg d\mu$$

By (50) and (52) we obtain

$$\left|\int_F sg d\mu\right| = \left|\int_{F^*} \hat{s}\hat{f} d\nu\right|$$

$$\le \epsilon \ \|\hat{s}\| \ \nu F^*$$

$$= \epsilon \ \|s\| \ \mu F.$$

Thus, $|g| \le \epsilon$ on F and by the assumption on its oscillation it now follows that $|g| \le 2\varepsilon$ on E. Since $\varepsilon > 0$ was arbitrary, the conclusion of the theorem follows. ■

<u>Remark</u>. A modified version of the main conclusion in Theorem 48 actually holds for functions in $C(M)$. Let μ be a measure on ∂G and let ν be a minimal measure with respect to μ. If $f \in C(M)$, then there exist closed subsets $\{E_n\}$ of ∂G with $\mu E_n \to 1$ such that for $z \in E_n$ and $\phi, \varphi \in M_z \cap \operatorname{spt}(\nu|_{E_n^*})$, we have

$$f(\phi) = f(\varphi).$$

<u>Proof</u>. If $f = \hat{g}$ for some $g \in H^\infty(G)$, then the conclusion holds by Theorem 48. A standard

argument extends the result to the case where $f = \sum_1^m \hat{g}_i \bar{\hat{h}}_i$ for some functions g_i and h_i in $H^\infty(G)$. The latter class of functions is uniformly dense in $C(\mathcal{M})$ by the Stone–Weierstrass theorem.

Let f be an arbitrary function in $C(\mathcal{M})$. By the reasoning of the previous paragraph there exists a sequence $\{f_m\}$ of functions in $C(\mathcal{M})$ that converge uniformly to f and satisfy our conclusion. Let F_m be a closed subset of ∂G with $\mu F_m > 1-(1/2^m)$ such that

$$\hat{f}_m(\phi) = \hat{f}_m(\varphi)$$

if ϕ, $\varphi \epsilon \mathcal{M}_z \cap \mathrm{spt}(\nu|_{F_m^*})$ for some $z \epsilon F_m$. Let $E_n = \bigcap_{m=n}^{\infty} F_m$. The conclusion follows easily.■

For any minimal measure ν and for any $f \epsilon H^\infty(G)$, let $\pi_\nu(f) = d\lambda/d\mu$ where λ is the unique measure on ∂G satisfying equation (47). We have that π_ν is a bounded linear transformation from $H^\infty(G)$ into $L^\infty(\mu)$ that satisfies $\pi_\nu(1)=1$. Furthermore, from Theorem 48, it now follows that $\pi_\nu(\chi)=\chi$. We must now show that π_ν is multiplicative.

To this end, let $f_1, f_2 \epsilon H^\infty(G)$ and let g_1, g_2, g_3 $\epsilon L^\infty(\mu)$ be such that

$$\pi_\nu(f_1)=g_1,\ \pi_\nu(f_2)=g_2, \text{ and } \pi_\nu(f_1 f_2)=g_3.$$

By Lusin's theorem there exist closed subsets E_n of ∂G such that

$$\mathrm{spt}(\mu|_{E_n}) = E_n,$$

$$g_i | E_n \text{ is continuous}$$

for all i and for all n, and

$$\lim_n \mu(E_n) = 1.$$

If $z \epsilon E_n$, then there exists

$$\alpha \epsilon \mathcal{M}_z(G) \cap \mathrm{spt}(\nu|_{E_n^*}),$$

and by Theorem 48

$$\hat{f}_1(\alpha) = g_1(z),$$

$$\hat{f}_2(\alpha) = g_2(z),$$

and

$$(f_1f_2)\hat{}(\alpha) = g_3(z).$$

Since the Gelfand transform is multiplicative, it follows that $g_3(z)=g_1(z)g_2(z)$. Consequently, $g_1g_2=g_3$ almost everywhere (μ), and the map π_ν is a unital representation of $H^\infty(G)$ into $L^\infty(\mu)$. Thus, we have proved the following theorem.

Theorem 53. Each minimal measure on $\mathcal{M}(G)$ induces a norm continuous unital representation of $H^\infty(G)$ into $L^\infty(\mu)$ that maps χ to χ.

We now show how Example 40 follows from this theorem and Theorem 48.

Details for Example 40. Suppose B is a Blaschke product whose zeros accumulate at every point on ∂D, and $f \in L^\infty(\mu)$ with $\|f\| \le 1$ where μ is any probability measure with spt $\mu \subset \partial D$. Let $\{E_n\}$ be a sequence of pairwise disjoint closed subsets of ∂D such that $\mu(\cup E_n)=1$, $\mathrm{spt}(\mu|_{E_n}) = E_n$ and $f|_{E_n}$ is continuous for all n.

The cluster set of B at each point of ∂D is $\bar{D}$, [25, p. 80]. Recall, if $f \in H^\infty(G)$ and $z \in \partial G$, then the cluster set of f at z, denoted C(f,z), is the set of all points $w \in \mathbb{C}$ such that there exists a sequence of points $\{z_n\}$ in G converging to z and $\lim_n f(z_n)=w$. Since $\hat{B}(\mathcal{M}_z)=C(B,z)$ (cf. [18,p.176]), it follows that the set

$$\eta_n \equiv \{\phi \in E_n^*: \hat{B}(\phi)=f(z) \text{ if } \phi \in \mathcal{M}_z\}$$

projects to all of E_n; i.e., $p(\eta_n)=E_n$. Since $\eta_n=\{\phi \in E_n^*: \hat{B}(\phi)=f(p\phi)\}$, it is a closed set and $C(E_n)$ can be considered as a subspace of $C(\eta_n)$. By the Hahn–Banach and Riesz representation theorems there exists a measure τ_n on η_n such that

$$\int_{E_n^*} \hat{h}\,d\tau_n = \int_{E_n} h\,d\mu$$

for every h in $C(E_n)$.

Let $\tau = \sum_n \tau_n$; clearly $\tau \in R$. Let ν be a minmal measure such that $\tau < \nu$. Using Theorem 48, we see clearly $\pi_\nu(B) = f$.■

Our next theorem is a converse to Theorem 53.

<u>Theorem 54</u>. Let π be a unital representation of $H^\infty(G)$ into $L^\infty(\mu)$ that maps χ to χ. Then there exists a minimal measure that induces π; i.e., $\pi = \pi_\nu$ for some minimal measure ν.

<u>Proof</u>. Fix a function f in $H^\infty(G)$. By Lusin's theorem there exists a countable collection $\{E_n\}$ of pairwise disjoint compact subsets of ∂G such that $\mathrm{spt}(\mu|_{E_n}) = E_n$, $(\pi f)|_{E_n}$ is continuous, and $\mu(\cup E_n) = 1$. Fix $z \in E_n$. The set of functions h in $L^\infty(\mu)$ such that

$$\operatorname*{ess\ lim}_{\substack{w \to z \\ w \in E_n}} h(w) = 0$$

is an ideal; hence, this set is contained in a maximal ideal ϕ for $L^\infty(\mu)$.
(Recall that

$$\operatorname*{ess\ lim}_{\substack{w \to z \\ w \in E_n}} h = \xi$$

means for each $\epsilon > 0$, there exists a neighborhood N of z such that

$$\mu(\{|f - \xi| < \epsilon\} \cap (N \cap E_n)) = \mu(N \cap E_n).)$$

If we identify ϕ with its corresponding multiplicative linear functional, then clearly $\phi(\chi) = z$. Now $\phi \circ \pi$ is a multiplicative linear functional on $H^\infty(G)$ and $\phi(\pi f) = (\pi f)(z)$.

Define, for each n, a subset ε_n of $\mathcal{M}$ by

$$\varepsilon_n \equiv \bigcup_{z \in E_n} \{\varphi \in \mathcal{M}_z : \hat{f}(\varphi) = (\pi f)(z)\}.$$

Clearly ε_n is closed and, from the last paragraph, $p(\varepsilon_n) = E_n$. Let τ_n be a positive measure on ε_n satisfying

$$\int_{\mathcal{E}_n} \hat{g}\, d\tau_n = \int_{E_n} g\, d\mu$$

for all $g \in C(E_n)$. Let $\tau_{\{f\}} = \Sigma \tau_n$. Note that $\tau_{\{f\}} \in R$.

Now fix a finite subset A of $H^\infty(G)$. We shall construct a measure τ_A in R just as we did above for $\{f\}$. Let $A = \{f_1, \ldots, f_n\}$. For each j, let $\{E_{jk}\}$ be the collection $\{E_n\}$ corresponding to f_j as done in the case for one function. Taking intersections of n sets in an appropriate manner, we obtain a countable collection $\{F_i\}$ of pairwise disjoint compact subsets of ∂G such that

$$\mu(E_{jk} \setminus \bigcup_{F_i \subset E_{jk}} F_i) = 0$$

for all j and k. Replace each F_i by $\operatorname{spt}(\mu \mid_{F_i})$.

Define for each i a subset $\mathcal{F}_i$ of $\mathcal{M}$ by

$$\mathcal{F}_i = \bigcup_{z \in F_i} \{\varphi \in \mathcal{M}_z : \text{for all } j,\ \hat{f}_j(\varphi) = (\pi f_j)(z)\}.$$

An argument, just like the one given in the case of a single function, shows that $p(\mathcal{F}_i) = F_i$. Define measures τ_i on $\mathcal{F}_i$ such that

$$\int_{\mathcal{F}_i} \hat{g}\, d\tau_i = \int_{F_i} g\, d\mu$$

for all $g \in C(F_i)$. Let $\tau_A = \Sigma \tau_i$ and note $\tau_A \in R$. In this way, for all finite subsets A of $H^\infty(G)$, we have obtained a measure τ_A in R.

Ordering the finite subsets of $H^\infty(G)$ by inclusion, we obtain a net $\{\tau_A\}$. Let τ be a cluster point of the net and let ν be a minimal measure such that $\tau < \nu$. We shall show that ν induces π.

Fix any function $f \in H^\infty(G)$ and one of its corresponding subsets E_n of ∂G described earlier. A subnet $\{\tau_\beta\}_{\beta \in B}$ of $\{\tau_A\}$ converges to τ. We may assume $f \in \beta$ for all

$\beta \in \mathcal{B}$. By Lemma 44, we have that $\tau_\beta |_{E_n^*} \to \tau |_{E_n^*}$. If $\{F_i\}$ is the sequence of subsets of ∂G for β in $\mathcal{B}$, then

$$\mu(E_n \setminus \bigcup_{F_i \subset E_n} F_i) = 0.$$

Moreover, for $F_i \subset E_n$ we have

$$\mathrm{spt}(\tau_\beta |_{F_i^*}) \subset \bigcup_{z \in F_i} \{\phi \in \mathcal{M}_z : (\pi f)(z) = \hat{f}(\phi)\}$$

$$\subset \bigcup_{z \in E_n} \{\phi \in \mathcal{M}_z : (\pi f)(z) = \hat{f}(\phi)\}$$

$$\equiv T.$$

Thus, taking the union over all $F_i \subset E_n$, we obtain

$$\mathrm{spt}(\tau_\beta |_{\cup F_i^*}) \subset T.$$

Let us now note that

$$\mathrm{spt}(\tau_\beta |_{E_n^*}) = \mathrm{spt}(\tau_\beta |_{\cup F_i^*}).$$

Using Lemmas 44 and 43, we obtain the fact that

$$\mathrm{spt}(\tau |_{E_n^*}) \subset T.$$

By the definition of ν we have

$$\mathrm{spt}(\nu |_{E_n^*}) \subset \mathrm{spt}\ (\tau |_{E_n^*}).$$

Thus,

$$(\pi f)(z) = \hat{f}(\phi)$$

for all $\phi \in \mathrm{spt}(\nu |_{E_n^*}) \cap \mathcal{M}_z$; which, in turn, implies from Theorem 48 that $\pi f = \pi_\nu(f)$. Since f was an arbitrary function, the proof is completed.∎

The next two lemmas show that the mapping $\nu \to \pi_\nu$ from minimal measures to representations is one-to-one.

<u>Lemma</u> 55. Nonequivalent minimal measures induce distinct representations.

<u>Proof</u>. Let σ and τ be nonequivalent minimal measures with respect to μ, our fixed

probability measure on ∂G. By definition there exists a compact subset E of ∂G such that $\mathrm{spt}(\mu|_E)=E$ and

$$\mathrm{spt}(\sigma|_{E^*}) \neq \mathrm{spt}(\tau|_{E^*}).$$

Without loss of generality, we may assume

$$\mathrm{spt}(\sigma|_{E^*}) \setminus \mathrm{spt}(\tau|_{E^*}) \neq \phi.$$

Let $\psi \in \mathrm{spt}(\sigma|_{E^*}) \setminus \mathrm{spt}(\tau|_{E^*})$ and let $\mathcal{O}$ be a basic open set in $\mathcal{M}$ s.t. $\psi \in \mathcal{O}$ and

$$\mathcal{O} \cap \mathrm{spt}(\tau|_{E^*}) = \phi.$$

By the definition of basic open sets for the weak-star topology, there exist $\epsilon>0$ and functions $f_1, \ldots, f_n$ in $H^\infty(G)$ such that $\hat{f}_i(\psi)=0$ for $i=1, \ldots, n$ and

$$\mathcal{O} = \{\varphi \in \mathcal{M}: |\hat{f}_i(\varphi)| < 2\epsilon \text{ for } i=1, \ldots, n\}.$$

Let

$$U = \{\phi \in \mathcal{M}: |\hat{f}_i(\phi)| < \varepsilon \text{ for } i=1,\ldots,n\}.$$

Since $\psi \in U$, it follows that $\sigma(E^* \cap U)>0$. Replacing E by a closed subset, we may assume that $(\pi_\sigma f_i)|_E$ and $(\pi_\tau f_i)|_E$ are continuous for each i. With this assumption we can no longer assume that $\psi \in \mathrm{spt}(\sigma|_{E^*})$.

As in the proof of Theorem 48, there exists a closed subset F of E such that $\mu F>0$ and $\sigma(F^* \setminus U)=0$. Since $\sigma(F^* \cap U)=\sigma(F^*)$, it follows that

$$\mathrm{spt}(\sigma|_{F^*}) \subset \bar{U}$$

$$\subset \{\phi \in \mathcal{M}: |\hat{f}_i(\phi)| \leq \epsilon \text{ for } i=1, \ldots, n\}.$$

Applying Theorem 48, we see that for $z \in E$ and for each i,

$$(\pi_\sigma f_i)(z) = \hat{f}_i(\phi)$$

for every ϕ in $\mathrm{spt}(\sigma|_{F^*}) \cap \mathcal{M}_z$, and

$$(\pi_\tau f_i)(z) = \hat{f}_i(\Theta)$$

for every Θ in $\mathrm{spt}(\tau|_{F^*})\cap\mathcal{M}_z$. From the last inclusion of the preceding paragraph it follows that for each i,

$$|\pi_\sigma f_i| \leq \epsilon \text{ a.e.}(\mu) \text{ on } F.$$

Since $\mathcal{O}\cap\mathrm{spt}(\tau|_{F^*})=\phi$, it follows that for each $\Theta\epsilon\,\mathrm{spt}(\tau|_{F^*})$ there exists an i such that $|\hat{f}_i(\Theta)|\geq 2\epsilon$. Hence, for each $z\in F$ there exists an i such that $|(\pi_\tau f_i)(z)|\geq 2\epsilon$. Since

$$F\subset \bigcup_{i=1}^{n} \{z\epsilon F\colon |(\pi_\tau f_i)(z)|\geq 2\epsilon\},$$

it follows that for some i, $|(\pi_\tau f_i)(z)|\geq 2\epsilon$ on a subset of F with positive μ measure. For that i, the functions $\pi_\tau f_i$ and $\pi_\sigma f_i$ are not equal, and the lemma is proved.■

<u>Lemma</u> 56. If two minimal measures are not equal, then they are not equivalent.

<u>Proof</u>. Let τ and ν be two minimal measures that are not equal. First, there exists a closed set $\varepsilon\subset\mathcal{M}$ such that

(57) $\nu(\varepsilon)<\tau(\varepsilon)\leq\mu(p\varepsilon)$.

Define measures β and γ on ∂G by setting

$$\beta A = \tau(A^*\cap\varepsilon)$$

and

$$\gamma A = \nu(A^*\cap\varepsilon)$$

for each measurable set $A\subset\partial G$. By Lemma 49 there exist measurable sets S and T such that $\beta=\chi_S\mu$ and $\gamma=\chi_T\mu$. By (57) there exists a closed set $U\subset S\setminus T$ with $\mu U>0$. Clearly $\mathrm{spt}(\tau|_{U^*})\subset\varepsilon$, but

$$\nu(\varepsilon\cap U^*) = 0, \text{ so}$$

$$\mathrm{spt}(\nu|_{U^*}) \not\subset \varepsilon.$$

Thus τ and ν are not equivalent.■

We may summarize the results obtained so far in this section as follows:

<u>Theorem</u> 58. The mapping $\nu\to\pi_\nu$ from the set of minimal measures to the set of

unital representations from $H^\infty(G)$ into $L^\infty(\mu)$ that send χ to χ is one-to-one and onto.

Proof. If ν is a minimal measure, then $[\nu]$ denotes the equivalence class in $\mathcal{R}$ that contains ν. Lemma 55 proves that the mapping $[\nu]\mapsto\pi_\nu$ is one-to-one. Lemma 56 shows that the equivalence class $[\nu]$ is a singleton for every minimal measure ν. That the mapping is onto follows from Theorem 54.∎

It is natural to ask, what is the spectral mapping theorem associated with a given unital representation π (from $H^\infty(G)$ into $L^\infty(\mu)$) that sends χ to χ? The next proposition answers this question.

Proposition 59. Let π be a representation from $H^\infty(G)$ into $L^\infty(\mu)$ that satisfies the conditions of Theorem 58. Let ν be the minimal measure inducing π. Then

$$\sigma(\pi(f)) = \hat{f}(\text{spt } \nu)$$

for all $f \in H^\infty(G)$. Note, by definition, if $g \in L^\infty(\mu)$, then $\sigma(g)$ is the μ essential range of g; i.e.,

$$\sigma(g)=\{\varsigma \in \mathbb{C}: \text{ for each neighborhood } \mathcal{O} \text{ of } \varsigma \text{ the set of } w \in \partial G \text{ such that } f(w) \in \mathcal{O} \text{ has positive } \mu \text{ measure}\}.$$

Proof. Let $\varsigma \in \sigma(\pi(f))$ and let $\epsilon>0$. There exists a compact set $K\subset\partial G$ such that $\mu(K)>0$, $\text{spt}(\mu|_K)=K$, $\pi(f)$ is continuous on K, and

$$|\pi(f)(w)-\varsigma| < \epsilon$$

for all $w \in K$. Using Theorem 48, we then see that $\pi(f)(w) \in \hat{f}(\text{spt } \nu)$ for every $w \in K$. Since $\hat{f}(\text{spt } \nu)$ is a closed set and ϵ was arbitrary, it follows that $\varsigma \in \hat{f}(\text{spt } \nu)$; whence $\sigma(\pi(f))\subset\hat{f}(\text{spt } \nu)$.

Now let $\varsigma \in \hat{f}(\text{spt } \nu)$ and let $\epsilon>0$. We let

$$U_\epsilon = \{\phi \in \text{spt } \nu: |\hat{f}(\phi)-\varsigma| \le \epsilon\}.$$

Since $\varsigma \in \hat{f}(\text{spt } \nu)$ and f is continuous on $\mathcal{M}$, $\nu(U_\epsilon)>0$. Therefore, $\mu(pU_\epsilon)>0$. Let E be a closed subset of pU_ϵ such that $\mu(E)>0$, $\text{spt}(\mu|_E)=E$ and $\pi(f)$ is continuous on E. It now follows, from Theorem 48, that

$$\mu(\{z \in E: \ |(f)(z)-\varsigma| \leq \epsilon\}) > 0.$$

Since ϵ was arbitrary, $\varsigma \in \sigma(\pi(f))$; therefore, $\sigma(\pi(f)) \supset \hat{f}(\text{spt } \nu)$ and the proof is completed.■

We have some unfinished business to attend to. Let us now complete the details to Theorem 42.

<u>Proof of Theorem 42</u>. Let ν be a minimal measure that induces π. Fix a point $z \in \partial D$. Let $\mathcal{F}$ be the set of functions f in $H^\infty(D)$ for which there exists a compact set E (which can depend on f) with full density at z such that $\pi(f)$ is continuous on E. (Recall that E has full density at z if

$$\lim_{r\to 0} \frac{m(E \cap \{w \in \partial D: \ |w-z|<r\})}{m(\{w \in \partial D: \ |w-z|<r\})} = 1.)$$

By Theorem 48 we have for all $f \in \mathcal{F}$

$$\pi(f)(z) = \hat{f}(\phi)$$

for all $\phi \in \text{spt}(\nu|_{E^*}) \cap \mathcal{M}_z(D)$.

We claim that

$$\phi \neq \bigcap_{f \in \mathcal{F}} \{\varphi \in \mathcal{M}_z(D): \hat{f}(\varphi) = \pi(f)(z)\}.$$

Suppose the claim is false. Since each set in the intersection is compact, there must exist a finite collection $\{f_1,\ldots,f_n\} \subset \mathcal{F}$ such that

$$\phi = \bigcap_{i=1}^{n} \{\varphi \in \mathcal{M}_z(D): \hat{f}_i(\varphi) = (\pi f_i)(z)\}.$$

Recall the proof of Theorem 54. Note from that argument we see that

$$\bigcap_{i=1}^{n} \{\phi \in \mathcal{M}_z(D): \ \hat{f}_i(\phi) = \pi(f_i)(z)\}$$

$$\supset \bigcap_{i=1}^{n} (\text{spt}(\nu|_{E_i^*}) \cap \mathcal{M}_z)$$

where $E_i = E(f_i)$. Let F be the intersection of the E_i's. The right-hand side of the inclusion above clearly contains the set

$$\text{spt}(\nu|_{F^*}) \cap \mathcal{M}_z.$$

In our construction of ν (in the proof of Theorem 54), we saw that this latter set is nonempty; note that

$$z \in \operatorname{spt}(m|_F)$$

because F has full density at z. This establishes the claim.

Let s(z) be a point in the intersection that was just proved to be nonempty. It follows easily now that

$$(\pi f)(z) = \hat{f}(s(z))$$

almost everywhere m for all $f \in H^\infty(D)$.■

Remark. The conclusion of Theorem 42 actually holds for every measure on ∂D. The modification of the proof is obvious with the following lemma.

Lemma 60. Let μ be a measure on ∂D and let E be a measurable subset of ∂D. Then almost every point of E has full density with respect to E.

Before proving the lemma, we should first define full density for a measure on ∂D. Let μ be a measure on ∂D and let E be a measurable subset of ∂D. For each $x \in \partial D$, let ℓ_x denote the collection of open subarcs of ∂D that are centered at x. The set E has full density at x if

$$\lim_{\substack{I \in \ell_x \\ mI \to 0}} \frac{\mu(E \cap I)}{\mu I} = 1.$$

The limit above may be replaced by lim inf.

Proof. Let F be the set of points in E that are of full density with respect to E. First, we shall show that F is measurable. Let A_k be the set of $x \in E$ for which there exists an arc I in ℓ_x such that $mI < 1/k$ and

$$\frac{\mu(E \cap I)}{\mu I} < 1 - (1/k).$$

Since each I in ℓ_x is an open arc, it follows that A_k is open in E. One easily sees that

$$E \setminus F = \bigcup_{j=1}^{\infty} \bigcap_{k=j}^{\infty} A_k,$$

and thus F is measurable.

Now we want to show that $\mu F = \mu E$. Suppose to the contrary that $E \setminus F$ has positive μ-measure. Then there exist $\epsilon > 0$ and a compact subset K of $E \setminus F$ with $\mu K > 0$ such that E has lower density less that $1-\epsilon$ at every point in K; that is, for each $x \in K$

$$(61) \quad \liminf_{I \in \ell_x,\ mI \to 0} \frac{\mu(E \cap I)}{\mu I} < 1 - \epsilon.$$

Let $\{H_n\}$ be a collection of open supersets of K such that $\mu H_n \to \mu K$. Fix n. For each $x \in K$ let I_x in ℓ_x be such that $I_x \subset H_n$ and equation 61 is satisfied with I_x in place of I. Since K is compact, there exist a finite number of these I_x's that cover K. Call this finite subcollection $\{I_j\}$. By [25, p.25] there exists a subcollection $\{J_i\}$ of the collection $\{I_j\}$ such that the J_i's are pairwise disjoint and

$$\sum_i \mu J_i \geq \tfrac{1}{2}\mu(\cup I_j) \geq \tfrac{1}{2}\mu K.$$

Let $U_n = \cup J_i$. Now we know that $U_n \subset H_n$ and $\mu U_n \geq \tfrac{1}{2}\mu K$; but if we compute using equation 61, then we obtain

$$(62) \quad \mu(U_n \cap E) < (1-\epsilon)\mu U_n.$$

Passing to a subsequence, if necessary, we may assume that $\mu U_n \to a$ where $a \geq (\tfrac{1}{2})\mu K$. Now

$$\mu(U_n \cap E) = \mu U_n - \mu(U_n \setminus E).$$

Since

$$\mu(U_n \setminus E) \leq \mu(H_n \setminus K).$$

and

$$\mu(H_n \setminus K) \to 0,$$

it follows that

$$\mu(U_n \cap E) \to a.$$

Letting $n \to \infty$ in equation 62, we obtain a contradiction.■

One drawback in studying representations from $H^\infty(D)$ into $L^\infty(m)$ from the viewpoint of these measurable cross sections is that many of them induce the same representation. The reader might be convinced of this fact (that many selector functions induce the same representation) from the proof of Theorem 42; but we shall illustrate this phenomenon through an example given in the next section (consult Example 76).

CHAPTER VI

REPRESENTATIONS OF $H^\infty(G)$ INTO $L^\infty(\mu)$ THAT ARE ISOMETRIES

We established in the last section that if G is a bounded domain in the plane and μ is any probability measure with spt $\mu \subset \partial G$, then there exist many unital representations π from $H^\infty(G)$ into $L^\infty(\mu)$ such that $\pi(\chi)=\chi$. In this section we focus our attention on the general problem of determining which of these representations are isometries; i.e., for which π does

$$\|\pi(f)\| = \|f\|$$

hold for all $f \in H^\infty(G)$?

(Our interest in this question was originally aroused by some operator theory problems. However, it seems to the authors that this is a very natural and basic function-theoretic issue. For if π is an isometric unital representation of $H^\infty(G)$ into $L^\infty(\mu)$, then for each $f \in H^\infty(G)$ it does justice, in a sense the reader can motivate for himself/herself, to call $\pi(f)$ the boundary values of f. Thus, in a real way, the study of this problem may be viewed as a study of the problem "what do boundary values mean in a domain?")

It turns out to be the case that there exist many domains G for which there <u>does not</u> exist a representation π from $H^\infty(G)$ into $L^\infty(\mu)$ that is an isometry, no matter what probability measure μ is specified on the boundary. The slit disc $G \equiv D \setminus \{\lambda \in \mathbb{R}: -1 \le \lambda \le 0\}$ is such a domain as we shall see.

<u>Question</u> 63. Which domains G have the property that they support some unital representation π from $H^\infty(G)$ into $L^\infty(\mu)$ that is an isometry for some measure μ with spt $\mu \subset \partial G$?

Unfortunately, we are unable to answer this question. The main result of this section is to answer (63) for the simply connected domains. Consult Theorem 94.

We begin by making some elementary observations. Suppose, at first, G is an arbitrary bounded domain, π is a unital representation from $H^\infty(G)$ into $L^\infty(\mu)$ where μ is a probability measure with spt $\mu\subset\partial G$, and ν is the minimal measure inducing π. We denote the Silov boundary of $H^\infty(G)$ by $\Upsilon(G)$. Note that if $g\in L^\infty(\mu)$, then $\|g\|$ equals the spectral radius of $\sigma(g)$. Therefore, it follows, from Proposition 59, that

$$(64)\quad \|\pi(f)\| = \sup_{\phi\in \mathrm{spt}\,\nu} |\hat{f}(\phi)|$$

for all $f\in H^\infty(G)$. Whence, π is an isometry if and only if spt $\nu\supset\Upsilon(G)$. Two interesting consequences of Equation 64 are the following.

Corollary 65. Let π: $H^\infty(G)\to L^\infty(\mu)$ be a unital representation induced by the minimal measure ν. Let E=spt ν. A necessary and sufficient condition that π has closed range is that $(H^\infty)^\wedge|_E$ is a closed subalgebra of C(E).

Proof. Suppose π has closed range and $\{f_n\}$ is a sequence of functions in $H^\infty(G)$ such that $\hat{f}_n\to g\in C(E)$ uniformly on E. From Equation 64 we see that $\{\pi(f_n)\}$ is a Cauchy sequence. Hence, there exists a function $h\in H^\infty(G)$ such that $\pi(f_n)\to\pi(h)$. Clearly (from Theorem 48), $g=\hat{h}|_E$.

Conversely, if $(H^\infty)^\wedge|_E$ is closed in C(E) and $\{f_n\}$ is a sequence of bounded analytic functions on G such that $\pi(f_n)\to h$ in $L^\infty(\mu)$, then by Equation 64 we see that $\{\hat{f}_n|_E\}$ is a Cauchy sequence. Hence, there exists a function $g\in H^\infty(G)$ such that $\hat{f}_n|_E\to\hat{g}|_E$. It follows from Theorem 48 that $h=\pi(g)$.∎

Corollary 66. Let π be a unital representation of $H^\infty(G)$ into $L^\infty(\mu)$. If π is one-to-one and has closed range, then π is an isometry.

Proof. Recall that if E is a closed subset of $\mathcal{M}(G)$, then the H^∞-convex hull of E, denoted $\hat{E}$, can be characterized as follows.

$$\hat{E} = \{\psi \in \mathcal{M}(G): |\hat{f}(\psi)| \le \|f\|_E \text{ for all } f \in H^\infty(G)\}$$

$$= \{\phi \in \mathcal{M}(G): \text{there exists a } c \ge 0 \text{ such that}$$

$$|\hat{f}(\phi)| \le c\|f\|_E \text{ for all } f \in H^\infty(G)\}.$$

(Consult [19, p. 39].) Clearly $\hat{E} = \mathcal{M}(G)$ if and only if $E \supset \Upsilon(G)$. Since the weak-star closure of G in $\mathcal{M}(G)$ is a boundary, it follows then that $E \supset \Upsilon(G)$ if $\hat{E} \supset G$.

Let $E = \text{spt } \nu$ where ν is the minimal measure inducing π. By the argument above and the remark after Equation 64, to show π is an isometry it is enough to show $\hat{E} \supset G$. Suppose not, let $\lambda \in G \setminus \hat{E}$. Then there exists a sequence $\{f_n\}$ of $H^\infty(G)$ functions such that $f_n(\lambda) = 1$ while $\hat{f}_n \to 0$ uniformly on E. Hence, for each n,

$$\frac{1-f_n}{\chi - \lambda} \in H^\infty(G)$$

and

$$\left[\frac{1-f_n}{\chi-\lambda}\right]^{\wedge} \to \left[\frac{1}{\chi-\lambda}\right]^{\wedge}$$

uniformly on E. Since π has closed range, there exists $g \in H^\infty(G)$ such that $\pi(g) = \frac{1}{\chi - \lambda}$.

Hence,

$$\pi(1-(\chi-\lambda)g) = 0.$$

But the kernel of π is zero; whence, $1 = (\chi - \lambda)g$, a blatant contradiction. ∎

We note that there are representations that are one-to-one and do not have closed range, and that have closed range and are not one-to-one. To get an easy example of the latter, fix $\lambda \in \partial G$ and let μ be the point mass measure at λ. Choose any $\phi_\lambda \in \mathcal{M}_\lambda(G)$ and define π from $H^\infty(G)$ into $L^\infty(\mu) = \mathbb{C}$ by

$$\pi(f)(\lambda) \equiv \phi_\lambda(f).$$

An example of the former will be presented in this section after we have discussed the properties of the natural representation ι_D of $H^\infty(D)$ into $H^\infty(m) \subset L^\infty(m)$. (Consult

Example 77. In this regard, also consult Theorem 114 in Section 7.) Recall that m is normalized Lebesgue measure on ∂D and $H^\infty(m)$ is the weak-star closure of the polynomials in $L^\infty(m)$. For each $f \in H^\infty(D)$, we let f^* be that function in $H^\infty(m)$ obtained from the radial boundary values of f; i.e.,

$$f^*(e^{i\Theta}) = \lim_{r \to 1} f(re^{i\Theta})$$

for almost all $\Theta \in [0, 2\pi)$. Clearly the map ι_D: $H^\infty(D) \to L^\infty(m)$ defined by $\iota_D(f) = f^*$ is an example of a unital representation.

One result that we shall prove is the following. (Consult the first and second paragraphs after Proposition 75.)

Theorem 67. If π is a unital representation of $H^\infty(D)$ into $L^\infty(m)$ and π is an isometry, then $\pi = \iota_D$.

With this last result we can give a more striking example of a unital representation that has closed range but is not an isometry.

Example 68. There exists a unital representation of $H^\infty(D)$ onto $L^\infty(m)$.

The construction of this representation depends on the following result. Recall from Section 4 that a sequence $\{z_n\}$ of points in D is interpolating if, for every bounded sequence $\{s_n\} \in \ell^\infty$, there exists $f \in H^\infty(D)$ such that $f(z_n) = s_n$.

Lemma 69. There exists an interpolating sequence of $H^\infty(D)$ with subsequences $B_1, B_2, \ldots$, such that B_i and B_j have no points in common for $i \neq j$ and $\bar{B}_i \supset \partial D$ for all i.

Proof. This is a routine exercise, but, for completeness, here is a solution. From [16, Theorem 9.2 and Theorem 9.1], we see that if

$$1 - |z_{n+1}| \leq 1/2(1 - |z_n|)$$

for all n, then $\{z_n\}$ is interpolating. Let $\{\varsigma_n\}$ denote a countable dense subset of ∂D (without repetitions). Let $a_n = 1 - 2^{-n}$. If we choose a sequence $\{z_n\}$ such that $|z_n| = a_n$, then $\{z_n\}$ will be interpolating.

First, relabel the a_n's by using a double index so that $\{a_n\}_{n=1}^{\infty}=\{a_{mn}\}_{m,n=1}^{\infty}$. For each m and n, let

$$b_{mn} = a_{mn}\varsigma_n.$$

Note that $\lim_{m\to\infty} b_{mn}=\varsigma_n$. Let $A_k=\{b_{kn}\}_{n=1}^{\infty}$. Now double index the A_k's and call them the A_{ij}'s. Let

$$B_i = \bigcup_{j=1}^{\infty} A_{ij}.$$

Now each B_i contains infinitely many of the A_k's and hence, for all n, infinitely many of the b_{kn}'s. Thus,

$$\varsigma_n \in \overline{B}_i$$

for all n. ■

Proof of Example 68. The collection of subsets of $L^{\infty}(m)$ that contain finitely many elements (and at least one) is a particularly ordered set under inclusion. For every such subset A of $L^{\infty}(m)$, we shall define a measure τ_A. We shall see that if τ is a minimal measure and τ is bigger than (>, the ordering given in Section 5) some weak-star cluster point of $\{\tau_A\}$, then

$$\pi_{\tau}(H^{\infty}(D)) = L^{\infty}(m).$$

(Of course, the notation π_{ν} means the representation induced by ν, a minimal measure.)

We start with the case where A consists of one function f. Let $\{E_n\}$ be a sequence of disjoint closed subsets of ∂D such that

(i) $m(\cup E_n) = 1$;

(ii) $f|_{E_n}$ is continuous;

(iii) $\operatorname{spt}(m|_{E_n}) = E_n$ for all n.

Let $H_n=\bigcup_1^n E_i$. Let g_n be a function in $C(\partial D)$ such that

$$g_n | H_n = f | H_n$$

and $\|g_n\| \leq \|f\|$ for each n.

From Lemma 69, there exist pairwise disjoint sequences $B_1, B_2, \ldots$, such that $\bar{B}_i \supset \partial D$ for every i and $\cup B_i$ is an interpolating sequence. By the definition of interpolating sequence, there exists $h \in H^\infty(D)$ such that for every n

(iv) $h(z) = \tilde{g}_n(z)$

if $z \in B_n$. Here we have used the notation $\bar{g}_n$ to denote the harmonic extension of g_n to the open disc D. (We note here that we will eventually show, after constructing the desired representation π, that $\pi(h)=f$.) Let C_n be the closure of B_n in the weak-star topology of $\mathcal{M}(D)$. If $\phi \in H_n^* \cap C_n$, then

$$\hat{h}(\phi) = f(\hat{\chi}(\phi)).$$

Why? Let $\{z_\alpha\}$ be a net in B_n such that

$$z_\alpha \to \phi \text{ weak-star.}$$

Then $\hat{\chi}(z_\alpha) \to \hat{\chi}(\phi)$, so $z_\alpha \to \hat{\chi}(\phi)$ and we already know that $\hat{\chi}(\phi) \in H_n$. The observation follows if we note that

$$h(z_\alpha) = \tilde{g}_n(z_\alpha) \text{ and}$$

(since $z_\alpha \in B_n$) that

$$\tilde{g}_n(z_\alpha) \to g_n(\hat{\chi}(\phi)) = f(\hat{\chi}(\phi)).$$

Since $\bar{B}_n \supset \partial D$, it follows easily that

$$p(C_n) \supset \partial D.$$

Thus,

$$p(C_n \cap E_n^*) = E_n.$$

Therefore, in the obvious manner, we can identify $C(E_n)$ with a closed subspace of

$C(C_n \cap E_n^*)$. By the Hahn–Banach and Riesz representation theorems, there exists a measure τ_n on C_n such that

$$\int \hat{u} d\tau_n = \int_{E_n} u dm$$

for all $u \in C(E_n)$. Clearly from this last equality, $\Sigma \tau_n \in R = R_m$. We now define τ_A as

$$\tau_A \equiv \Sigma \tau_n.$$

Let us now consider the (arbitrary) case where A is a finite subset of $L^\infty(m)$, say $A = \{f_1, \ldots, f_q\}$. For each $f_i \in A$, we let $\{E_{ni}\}_{n=1}^\infty$ be the collection $\{E_n\}$ corresponding to f_i as done in the case for one function. By taking the appropriate intersections of the E_{ni}'s, we obtain a collection $\{F_m\}$ of closed subsets of ∂D satisfying

(v) $f_i|_{F_m}$ is continuous for all i and for all m;

(vi) $F_m \cap F_k = \phi$ whenever $m \neq k$;

(vii) $m(E_{ni} \setminus \bigcup_{F_m \subset E_{ni}} F_m) = 0$ for all i and for all n.

(It follows then from (vii) that $m(\cup F_m) = 1$.) Also, if $F_m \subset E_{ni}$, we may assume that $n \le m$. Furthermore, by discarding some sets of measure zero, we may assume that $\operatorname{spt}(m|_{F_m}) = F_m$. (It may now happen that $F_m = \phi$ for some m.) For each i, $1 \le i \le q$, let h_i be a function in $H^\infty(D)$ satisfying (iv). Note, the sequences $B_1, B_2, \ldots$, specified in this case are independent of the given f_i.

As done in the case where A is a singleton, we let τ_m be a measure on $C_m \cap F_m^*$ satisfying

$$\int \hat{u} d\tau_m = \int_{F_m} u dm$$

for all $u \in C(F_m)$. We now define τ_A by

$$\tau_A \equiv \Sigma\, \tau_m,$$

Let us now verify

(viii) $\operatorname{spt}(\tau_A | F_n^*) \subset \{\phi \in F_m^*: \hat{h}_i(\phi) = f_i(\hat{\chi}(\phi))\}$

for all m and for all i, $1 \leq i \leq q$. First of all, we fix m and note

$$\operatorname{spt}(\tau_A | F_m^*) = \operatorname{spt}(\tau_m)$$

because $\tau_A(F_m^*) = \tau_m(F_m^*)$. Now, by construction, spt $\tau_m \subset C_m \cap F_m^*$. Fix i, $1 \leq i \leq q$. By construction there exists $n \leq m$ such that

$$F_m \subset E_{ni}.$$

From our observation in the first case (where A is a singleton) if

$$\phi \in \left(\bigcup_{j=1}^{m} E_{ji}^*\right) \cap C_m,$$

then

$$\hat{h}_i(\phi) = f_i(\hat{\chi}(\phi)).$$

Since $\bigcup_{n=1}^{m} E_{ni}^* \supset F_m^*$, it now follows that if $\phi \in C_m \cap F_m^*$, then

$$\hat{h}_i(\phi) = f_i(\hat{\chi}(\phi)).$$

One now combines the facts established to get the result of (viii).

We now establish a fact that implies (viii). (We use (viii) to prove this fact.) We want to show for all i, $1 \leq i \leq q$,

(ix) $\operatorname{spt}(\tau_A | E_{ni}^*) \subset \{\phi \in E_{ni}^*: \hat{h}_i(\phi) = f_i(\hat{\chi}(\phi))\}$

for any n. To see this, fix any n and i; from (viii) we see

$$\bigcup_{F_m \subset E_{ni}} \operatorname{spt}(\tau_A | F_m^*) \subset \{\phi \in E_{n_i}^*: \hat{h}_i(\phi) = f_i(\hat{\chi}(\phi))\}.$$

Recalling (vii), we see

$$\tau_A\left[\bigcup_{F_m \subset E_{ni}} \operatorname{spt}(\tau_A | F_m^*)\right] = \tau_A(E_n^*);$$

so we see that

$$\{\phi \in E_{ni}^*: \hat{h}_i(\phi) = f_i(\hat{\chi}(\phi))\}$$

is a closed set containing a carrier of $\tau_A | E_{ni}^*$. Hence, the inclusion given in (ix) follows.

Let τ be a weak-star cluster point of the net $\{\tau_A\}$ (A runs over all nonempty finite subsets of $L^\infty(m)$). Obviously $\tau \in R=R_m$. Let σ be a minimal measure such that $\sigma > \tau$. Let π be the representation induced from σ. We now show $\pi(H^\infty(D)) = L^\infty(m)$.

Fix $f \in L^\infty(m)$ and let $\{E_n\}$ denote that sequence of pairwise disjoint closed subsets of ∂D specified earlier for the case $A=\{f\}$. Let h be a function in $H^\infty(D)$ satisfying (iv) for this case ($A=\{f\}$). We now show $\pi(h)=f$.

Let $\{\tau_\alpha\}$ be subnet of $\{\tau_A\}$ that converge to τ. By the defintion of subnet there exists α_0 such that whenever $\alpha > \alpha_0$, we have $f \in \alpha$. Since $\{\tau_\alpha\}_{\alpha>\alpha_0}$ also converges to τ and, for all $\alpha > \alpha_0$,

$$\operatorname{spt}(\tau_\alpha | E_n^*) \subset \{\phi \in E_n^*: \hat{h}(\phi) = f(\hat{\chi}(\phi))\}$$

for all n (see (ix)), it follows from Lemmas 44 and 43, in that order, that

$$\operatorname{spt}(\tau | E_n^*) \subset \{\phi \in E_n^*: \hat{h}(\phi) = f(\hat{\chi}(\phi))\}$$

for all n. One now applies Theorem 48 to get the result $\pi(h)=f$, if one just observes that

$$\operatorname{spt}(\sigma | E_n^*) \subset \operatorname{spt}(\tau | E_n^*)$$

because $\sigma > \tau$. ∎

Using Theorem 67 (whose proof you hold our indenture for) and Corollary 63, we see that the kernel of the representation π given in the proof of Example 68 is nontrivial. We note, though, that

$$\ker \pi \supset \{f \in H^\infty(D): f(z)=0 \text{ for all } z \in \bigcup_i B_i\};$$

a fact easily established from the details of the example. (The B_i's are those sets specified in the details.) Before we return to the main topic of this section, we point out some curious results that follow from Example 68.

Corollary 70. Let π be a unital representation of $H^{\infty}(D)$ onto $L^{\infty}(m)$. Then the natural representation (induced from π) of

$$\frac{H^{\infty}}{\mathrm{Ker}\,\pi} \text{ onto } L^{\infty}(m)$$

is a Banach algebra isomorphism that is the identity on polynomials.

Corollary 71. Let π be a unital representation of $H^{\infty}(D)$ onto $L^{\infty}(m)$. Let ν be the minimal measure that induces π. Let Γ denote the Banach algebra homomorphism from $H^{\infty}(D)$ into $C(\mathrm{spt}\ \nu)$ given by

$$\Gamma(f) \equiv \hat{f}\,|_{\mathrm{spt}\ \nu}$$

for each $f \in H^{\infty}(D)$. It follows that Γ maps $H^{\infty}(D)$ onto $C(\mathrm{spt}\ \nu)$.

Proof. From Corollary 65 we have Γ has closed range. Note that if $\pi(h_1)=f$ and $\pi(h_2)=f$, then

$$(\hat{h}_1\,|_{\mathrm{spt}\ \nu}) = (\hat{h}_2\,|_{\mathrm{spt}\ \nu}).$$

(The last statement follows easily from Theorem 48.) An application of the Stone–Weierstrass theorem now yields the result. ∎

Proposition 72. Let π be a unital representation of $H^{\infty}(D)$ onto $L^{\infty}(m)$. If ν is the minimal measure that induces π, then

$$C(\mathrm{spt}\ \nu) = L^{\infty}(\nu).$$

Proof. Let $g \in L^{\infty}(\nu)$; we shall find a function $h \in H^{\infty}(D)$ such that $\hat{h}\,|_{\mathrm{spt}\ \nu}=g$ almost everywhere ν. Let $\{E_n\}$ be an increasing sequence of compact subsets of spt ν such that $\nu(E_n)\to 1$ and $g\,|_{E_n}$ is continuous for all n. For each n, define a measure τ_n on ∂D by

$$\tau_n(F) \equiv \nu(F^* \cap E_n)$$

for every measurable set F. By Lemma 49 we see there exists a measurable set A_n such that

$$\tau_n = \chi_{A_n} m$$

for each n. Clearly, $m(A_n)\to 1$. Since $\nu(A_n^*)=\nu(A_n^*\cap E_n)$, one sees

$$\operatorname{spt}(\nu|_{B^*}) \subset E_n$$

for all closed subsets B of A_n. Choose a sequence $\{B_n\}$ of closed subsets of ∂D such that $B_n\subset A_n$ for all n and $m(B_n)\to 1$. (Note $\{A_n\}$ is an increasing sequence of sets.)

Now $g|_{B_n^*\cap E_n}$ is continuous; so, from Corollary 68 and Tietze's extension theorem, there exists a function $h_n\in H^\infty(D)$ such that

$$\hat{h}_n|_{B_n^*\cap E_n} = g|_{B_n^*\cap E_n}$$

for every n. For each n, we now choose a closed subset $C_n\subset B_n$ such that $(\pi h_n)|_{C_n}$ is continuous and so that $m(C_n)\to 1$.

There exists $f_n\in C(C_n)$ such that

$$g|_{C_n^*\cap E_n} = \hat{f}_n|_{C_n^*\cap E_n}$$

Thus, there exists a function f in $L^\infty(m)$ such that

$$f(p\psi) = g(\psi)$$

almost everywhere on spt ν. Choose a function $h\in H^\infty(D)$ such that

$$\pi h = f.$$

The reader can easily check that $\hat{h}=g$ almost everywhere ν. ■

We now return to the principal topic of this section. Our primary task now shall be the verification of Theorem 67. We shall identify in the usual fashion the maximal ideal space of $L^\infty(m)$ with $\Upsilon(D)$. Consult [28, p. 174]. Hence, we view the Gelfand transform of $L^\infty(m)$ as a *-isometric isomorphism of $L^\infty(m)$ onto $C(\Upsilon(D))$. By the Riesz representation theorem, there is a unique regular Borel measure $\hat{m}$ on $\Upsilon(D)$ such that

$$(73)\quad \int_{\partial D} f\,dm = \int_{\Upsilon(D)} \hat{f}\,d\hat{m}$$

for every $f \in L^{\infty}(m)$. We will use at will the results in [19, Chapter 1, Section 9]. In particular, we note that $\hat{m}$ is a normal measure with spt $\hat{m} = \Upsilon(D)$. Note then that $\hat{m}(F) = 0$ for every nowhere dense measurable subset of $\Upsilon(D)$.

It follows from equation 73 and the definition of the set R_{m}, that $\hat{m} \in R_{m}$. We now want to show $\hat{m}$ is the minimal measure inducing the representation ι_D (recall that this (natural) representation is defined in the paragraph preceding the statement of Theorem 67). But first we present two propositions.

<u>Propostion</u> 74. Let π be a unital representation from $H^{\infty}(D)$ into $L^{\infty}(m)$. If π is an isometry and ν is the minimal measure that induces π, then spt $\nu = \Upsilon(D)$.

<u>Proof</u>. Looking at the sentence after equation (64), we see that spt $\nu \supset \Upsilon(D)$. To prove the reverse inclusion, suppose there exists a $\phi \in (\text{spt } \nu) \setminus \Upsilon(D)$. Since $\mathcal{M}(D)$ is compact (hence, completely regular), there exist disjoint open sets U and V such that $\phi \in U$ and $V \supset \Upsilon(D)$.

For all Borel sets $E \subset \partial D$ define

$$\beta(E) = \nu(\bar{U} \cap E^{*}).$$

Clearly, β is a measure on ∂D; in fact, by Lemma 49 we have

$$\beta = \chi_A m$$

for some measurable subset $A \subset \partial D$. Since $\phi \in \text{spt } \nu$, clearly $\beta \neq 0$; therefore, there exists a compact set $K \subset A$ such that $m(K) > 0$. It follows easily then that

$$\text{spt } \nu \subset \overline{(\partial D \setminus K)^{*}} \cup \bar{U}.$$

We claim that this last inclusion is a contradiction to the fact that spt $\nu \supset \Upsilon(D)$.

To see this, choose a function f in $H^{\infty}(D)$ satisfying

$$|f| = \begin{cases} 1 & \text{on } \partial D \setminus K \\ 2 & \text{on } K. \end{cases}$$

Using the reflection principle, we may extend f to be an analytic function across $\partial D \setminus K$. Thus, $|\hat{f}| = 1$ on $(\partial D \setminus K)^{*}$; consequently, $|\hat{f}| = 1$ on $\overline{(\partial D \setminus K)^{*}}$. But $\|f\| = 2$ because

$m(K)>0$. It follows then that $\overline{\partial D \setminus K)}^{*}$ cannot contain $K^{*} \cap \Upsilon(D)$; otherwise, we would then have

$$\overline{(\partial D \setminus K)}^{*} \supset \Upsilon(D)$$

which is a blatant contradiction to the fact $\Upsilon(D)$ is a boundary. Since $\bar{U} \cap \Upsilon(D)=\phi$, we easily see spt $\nu \supsetneq \Upsilon(D)$. Therefore, the conslusion of the propostion follows. ■

Proposition 75. Let ν be any positive measure on $\mathcal{M}(D)$ such that spt $\nu=\Upsilon(D)$. Then $\hat{m} << \nu$.

Proof. Suppose $\nu(K)=0$ for some compact subset $K \subset \Upsilon(D)$. We observe that K is nowhere dense because spt $\nu=\Upsilon(D)$. Hence, $\hat{m}(K)=0$ and the result immediately follows. ■

Let ν be the minimal measure that induces an isometric unital representation of $H^{\infty}(D)$ into $L^{\infty}(m)$. By the last two propositions and the fact $\hat{m} \in R_m$, one sees that $\nu=\hat{m}$. Hence, from Theorem 58 there is at most one such representation. But ι_D is such a representation. We have paid one of our debts. We have proved:

Theorem 67. Let π be a unital representation from $H^{\infty}(D)$ into $L^{\infty}(m)$. If π is an isometry, then $\pi=\iota_D$ and the minimal measure inducing π is $\hat{m}$.

Before we begin our work on determining the simply connected regions that support isometric unital representations, we finish paying our bills.

Example 76. We now exhibit two different measurable cross sections that induce ι_D. (Consult the last paragraph of Section 5 and Theorem 42. Also recall that here s measurable means only that $\hat{f} \circ s$ is measurable for each f in $H^{\infty}(D)$.) To specify the first selector function s_1, recall the proof of Theorem 42. From that argument and the fact that spt $\hat{m} = \Upsilon(D)$ we see that

$$\phi \neq \bigcap_{f \in \mathcal{F}} \{\psi \in \Upsilon(D) \colon \hat{f}(\psi) = \iota_D(f)(z)\}.$$

Therefore, if $s_1(z) \in \Upsilon(D)$ is a point that belongs to this intersection, then

$$\iota_D(f)(z) = \hat{f}(s_1(z))$$

almost everywhere m on ∂D, for all $f \in H^\infty(D)$.

For a second selector function, fix $z \in \partial D$. Note $\lim_{n\to\infty}(1-\frac{1}{n^2})z = z$. Pick a cluster point $s_2(z) \in \mathcal{M}_z(D)$ of the sequence $\{(1-\frac{1}{n^2})z\}_n$. (We have imbedded D into $\mathcal{M}(D)$.) The point $s_2(z) \notin \Upsilon(D)$ because there exists a Blaschke product B such that $B((1-\frac{1}{n^2})z)=0$ for all n (hence, $\hat{B}(s_2(z))=0$); but, (cf. [28, p. 179]) all Blaschke products are unimodular on $\Upsilon(D)$. Now $\hat{f}\circ s_2 = \iota_D(f)$ if $f \in H^\infty(D)$ because

$$\lim_{n\to\infty} f((1-\frac{1}{n^2})z) = f(z) \text{ almost everywhere m on } \partial D. \blacksquare$$

<u>Example</u> 77. We now exhibit a unital representation π from $H^\infty(D)$ into $L^\infty(m)$ that is one-to-one and does not have closed range. (Consult the paragraph after the proof of Corollary 66.) Let ν be a minimal measure in R_m such that $\nu \neq \hat{m}$. Hence, there exists a closed subset E of ∂D satisfying

$$0 < m(E) < 1,$$

and

$$\nu|_{E^*} \neq \hat{m}|_{E^*}.$$

We leave it to the reader to check that

$$\tau \equiv \nu|_{E^*} + \hat{m}|_{(\partial D \setminus E)^*}$$

is a minimal measure in R_m and $\tau \neq \hat{m}$. Let π_τ be the representation induced by τ. From Theorem 48 and Theorem 67 we see that

$$\pi_\tau(f) = f^*$$

almost everywhere $m|_{\partial D\setminus E}$. Using the fact that if $f \in H^\infty(D)$ and if f^* vanishes on a set of positive (m) measure, then $f^*=0$ almost everywhere, we see that π_τ is one-to-one. (Note, $m(\partial D \setminus E)>0$.) Since $\pi_\tau \neq \iota_D$ (because $\tau \neq \hat{m}$), we get from Theorem 67 that π_τ is not an isometry; whence, from Corollary 66 we see π_τ does not have closed range. (For more results on the problem of when π is one-to-one, consult Theorem 114.) ∎

We now devote our attention to the problem of answering Question 63 if we restrict ourselves to the simply connected case. <u>For the rest of this section</u>, unless we explicitly state otherwise, <u>we shall assume that</u> G <u>is a bounded simply connected domain</u>. Our objective now is to show that if there is a unital representation π from $H^\infty(G)$ into $L^\infty(\mu)$ (where spt $\mu \subset \partial G$) that is an isometry, then there is a representation that is an isometry where the measure on the boundary is harmonic measure. But, first we need to establish some notation and state some well-known results. The following is a synopsis of section 1 in [13].

Since ∂G is connected, every point on ∂G is regular for the Dirichlet problem in G; i.e., if $f \in C_r(\partial G)$, then f has a unique continuous extension to $\bar{G}$ which is harmonic on G; we denote its restriction to G by $\tilde{f}$. Then for $z \in G$,

$$\tilde{f}(z) = \int_{\partial G} f d\lambda_z$$

where λ_z is harmonic measure for z with respect to G, a probability measure on ∂G. For $w \in D$, we denote by m_w the harmonic measure (in this case the Poisson kernel) for w with respect to D. Hence, $m=m_0$ in this case. It is well-known that if $z_1, z_2 \in G$, then for some $M>0$, $M^{-1}\lambda_{z_1} \leq \lambda_{z_2} \leq M\lambda_{z_1}$, so that λ_{z_1} and λ_{z_2} define isomomorphic L^P spaces. It is easy to see that if $g \in L^1(\lambda_{z_0})$ where $z_0 \in G$ and if we define a function $\tilde{g}$ on G by

$$\tilde{g}(z) = \int_{\partial G} g d\lambda_z,$$

then $\tilde{g}$ is harmonic in G.

Now let ψ be a conformal map of D onto G. Then ψ has radial boundary values almost everywhere m on ∂D; i.e.,

$$\psi(re^{i\Theta}) \to \psi^*(e^{i\Theta})$$

say, as $r \to 1$, for almost all $\Theta \in [0,2\pi)$. Using Lusin's theorem, we can find an increasing sequence $\{E_n\}$ of compact subsets of ∂D such that $\partial D \setminus \cup E_n$ has measure zero, ψ^* is defined on $E \equiv \cup E_n$, and $\psi^* |_{E_n}$ is continuous for each n.

One can show that if $w \in D$ and $z = \psi(w)$, then

$$(78) \quad \int_{\partial G} f d\lambda_z = \int_{\partial D} (f \circ \psi^*) dm_w$$

for all $f \in L^1(\lambda_z)$. Furthermore,

$$(79) \quad \lambda_z(F) = m_w(\psi^{*-1}(F))$$

for all measurable subsets F of ∂G. It follows easily then that if we write $Tf = f \circ \psi^*$, then T is an isometric isomorphism of $L^P(\lambda_z)$ into $L^P(m_w)$ for any p with $1 \leq p \leq \infty$.

We say that G is nicely connected if there is a measurable set $S \subset \partial D$ of measure zero such that ψ^* is one-to-one on $\partial D \setminus S$. In this case the set E can be constructed so that $E \cap S = \phi$, and then ψ^* maps each E_n homeomorphically onto $\psi^*(E_n)$. Thus, ψ^* has a measurable inverse defined on $\psi^*(E)$ which implies T maps $L^P(\lambda_z)$ onto $L^P(m_w)$ for all p.

Given the natural unital representation ι_D of $H^\infty(D)$ into $L^\infty(m)$, we construct a natural unital representation ι_G of $H^\infty(G)$ into $L^\infty(\lambda_z)$ for any nicely connected domain G in the obvious way; if $z = \psi(0)$, then

$$(80) \quad \iota_G(g) = T^{-1}((g \circ \psi)^*)$$

for any $g \in H^\infty(G)$. That is to say,

$$\iota_G(g) = T^{-1}(\iota_D(g \circ \psi))$$

for all $g \in H^\infty(G)$. One easily checks that ι_G is a unital representation that is an isometry (clearly $\iota_G(1)=1$ and $\iota_G(\chi)=\chi$). We leave it to the reader to check that ι_G is independent of the base point $z=\psi(0)$ that we choose. It is well-known that ι_D is a weak-star, weak-star continuous mapping. It follows then that ι_G is a weak-star homeomorphism for all nicely connected domains G.

Lemma 81. (Let G be a bounded simply connected domain.) If there exist a probability measure μ with spt $\mu \subset \partial G$ and a unital representation from $H^\infty(G)$ into $L^\infty(\mu)$ that is an isometry, then $\lambda_z << \mu$ for all $z \in G$.

Proof. Let π: $H^\infty(G) \to L^\infty(\mu)$ be the representation that is an isometry and let ν denote the minimal measure that induces π.

If the conclusion was false, then there exists a compact set $E \subset \partial G$ such that $\mu(E)=0$ and $\lambda_z(E)>0$. Let $u=(\chi_E - 1)$ and extend u to $\bar{G}$ by defining

$$u = \begin{cases} \tilde{u} \text{ on } G \\ u \text{ on } \partial G \end{cases}$$

Since $\partial G \setminus E$ is open and every point on ∂G is regular, one sees that u is continuous at every point in $\partial G \setminus E$. Since G is simply connected, $\tilde{u}$ has a harmonic conjugate v defined on G. Define a function f in $H^\infty(G)$ by letting

$$f = \exp(\tilde{u}+iv).$$

Since $|f| = e^{\tilde{u}}$ on G and $\lambda_z(E)>0$, it follows that $\|f\|=1$. Hence, $\|\pi(f)\|=1$.

Recalling that u is continuous at every point in $\partial G \setminus E$, we see that

(82) $C(|f|,\alpha) = \{e^{-1}\}$

for every $\alpha \in \partial G \setminus E$. (Recall, from Section 4 in the paragraph preceding Corollary 36, that $C(f,\alpha)$ denotes the cluster set of f at α. This set was defined only for functions in $H^\infty(G)$, but clearly it can be defined for any bounded function defined on G.) Noting that ν is carried by $\mathcal{M} \setminus E^*$ and recalling the fact that

$$C(f,\alpha) = \hat{f}(\mathcal{M}_\lambda)$$

for every $\lambda \in \bar{G}$, we see from Thereom 48 and Equation (82) that

$$|\pi(f)(\alpha)| = e^{-1}$$

μ almost everywhere; whence, $\|\pi(f)\| = e^{-1}$, contradicting the fact that $\|\pi(f)\| = 1$. ∎

If π: $H^\infty(G) \to L^\infty(\mu)$ is a unital representation that is an isometry, then, by the last lemma, we can write $\mu = \mu_1 + \mu_2$ where μ_1 and λ_z (for any $z \in G$) are mutually absolutely continuous measures, $\mu_1 \perp \mu_2$ and $\mu_2 \perp \lambda_z$. Without loss of generality, one may assume then that $\mu_1 = \lambda_z$. Writing $L^\infty(\mu) = L^\infty(\lambda_z) \oplus L^\infty(\mu_2)$ in the obvious way, we see that there exist unital representations π_1 and π_2 from $H^\infty(G)$ into $L^\infty(\lambda_z)$ and $L^\infty(\mu_2)$, respectively, such that $\pi = \pi_1 \oplus \pi_2$. Our next goal is to show that the representation π_1 must be an isometry. But first some preliminaries.

It is easy to see that if μ_1 and μ_2 are probability measures on ∂G and $\mu_1 \perp \mu_2$, then for every measure $\tau_1 \in R_{\mu_1}$ and for every measure $\tau_2 \in R_{\mu_2}$, it follows that

$$\tau_1 \perp \tau_2.$$

(Reason: If E is a measurable subset of ∂G such that $\mu_1(E) = \mu_2(\partial G \setminus E) = 0$, then one sees that $\tau_1(E^*) = \tau_2(\mathcal{M} \setminus E^*) = 0$.)

We now describe a measure in R_{λ_z}, where $z \in G$, that is central to the results in this section. We fix a point $z_0 \in G$.

As before, we let ψ be the conformal map of D onto G such that $\psi(0) = z_0$. The map ψ then induces a natural isometric isomorphism T_ψ of $H^\infty(G)$ onto $H^\infty(D)$ defined by

$$T_\psi(f) \equiv f \circ \psi$$

for all $f \in H^\infty(G)$. Hence, T_ψ^*, the Banach space adjoint of T_ψ, is an isometric isomorphism of $H^\infty(D)^*$ onto $H^\infty(G)^*$. Note for each $\phi \in H^\infty(D)^*$ and for each $f \in H^\infty(G)$,

$$(T_\psi^* \phi)(f) = \phi(T_\psi f)$$

$$= \phi(f \circ \psi).$$

One easily checks that $T_\psi^*(\mathcal{M}(D)) = \mathcal{M}(G)$. We let $h_\psi \equiv T_\psi^* \mid \mathcal{M}(D)$.

If $\phi_\alpha \to \phi$ weak-star in $\mathcal{M}(D)$, then clearly, $\phi_\alpha(f \circ \psi) \to \phi(f \circ \psi)$ for every $f \in H^\infty(G)$. Hence, $h_\psi(\phi_\alpha) \to h_\psi(\phi)$ weak-star in $\mathcal{M}(G)$. Thus, h_ψ is a weak-star homeomorphism of $\mathcal{M}(D)$ onto $\mathcal{M}(G)$. An elementary computation shows that

$$h_\psi(\Upsilon(D)) = \Upsilon(G).$$

Now, for notational convenience, we suppress a subscript and let $\lambda = \lambda_{z_0}$. We define a normal measure $\hat{\lambda}$ on $\Upsilon(G)$ by setting

$$\hat{\lambda} = \hat{m} \circ h_\psi^{-1};$$

i.e.,

$$\hat{\lambda}(F) = \hat{m}(h_\psi^{-1}(F))$$

for all Borel subsets $F \subset \Upsilon(G)$. Clearly, spt $\hat{\lambda} = \Upsilon(G)$.

We claim that $\hat{\lambda} \in R_\lambda$. (This is a well-known fact, but, for completeness, we present the details.) Observing that $\hat{\psi}^* = \hat{\psi} \mid \Upsilon(D)$ and $\hat{\psi}(\Upsilon(D)) \subset \partial G$, we see that if f is a polynomial in z and $\bar{z}$,

$$(f \circ \psi^*)\hat{} = f \circ (\psi^*)\hat{}$$

$$= f \circ \hat{\psi};$$

therefore, from the Stone-Weierstrauss theorem, this equality holds for every $f \in C(\partial G)$. Also, for each $f \in C(\partial G)$ and $\phi \in \Upsilon(D)$ we have

$$\hat{f}(h_\psi(\phi)) = f(\hat{\chi}(h_\psi(\phi)))$$

$$= f(\phi(\chi \circ \psi))$$

$$= f(\phi(\psi))$$

$$= f(\hat{\psi}(\phi)).$$

Thus, for $f \in C(\partial G)$

$$\int_{\partial G} f\,d\lambda = \int_{\partial D} f \circ \psi^{*} dm$$

$$= \int_{\Upsilon(D)} (f \circ \psi^{*})\hat{}\, d\hat{m}$$

$$= \int_{\Upsilon(D)} f \circ \hat{\psi} d\hat{m}$$

$$= \int_{\Upsilon(D)} \hat{f} \circ h_{\psi} d\hat{m}$$

$$= \int_{\Upsilon(G)} \hat{f} d\hat{\lambda}$$

That is, for every $f \in C(\partial G)$

$$\int_{\partial G} f\,d\lambda = \int_{\Upsilon(G)} \hat{f} d\hat{\lambda};$$

the claim now follows.

Perhaps it is worth mentioning here that, there exists an isometry π: $H^{\infty}(G) \to L^{\infty}(\mu)$ if and only if $\hat{\lambda}$ is a minimal measure. The proof of this fact will come later. We now have developed the machinery to prove the following (what we called earlier our next objective).

<u>Lemma</u> 83. (Let G be a bounded simply connected domain.) Fix any $z_0 \in G$. There exist a probability measure μ with spt $\mu \subset \partial G$ and a unital representation from $H^{\infty}(G)$ into $L^{\infty}(\mu)$ that is an isometry if and only if there exists a unital representation from $H^{\infty}(G)$ into $L^{\infty}(\lambda_{z_0})$ that is an isometry.

<u>Proof</u>. Suppose π: $H^{\infty}(G) \to L^{\infty}(\mu)$ is a unital representation that is an isometry. Using the notation and decompositions set forth in the paragraph after the proof of Lemma 81, we shall show that π_1 is an isometry. Again let $\lambda = \lambda_{z_0}$.

Let ν be the minimal measure inducing π and let ν_i be the minimal measure inducing π_i for i equal 1 and 2. Note that $\nu = \nu_1 + \nu_2$ and, as we saw earlier, $\nu_1 \perp \nu_2$. We also have $\nu_2 \perp \hat{\lambda}$ because $\hat{\lambda} \in R_\lambda$.

Suppose that π_1 is not an isometry. We shall arrive at a contradiction. (Hence, the lemma will be established.) There exists a $g \in H^\infty(G)$ such that $\|g\| > \|\pi_1(g)\|$. Since $\Upsilon(G)$ is a boundary for $H^\infty(G)$, there exists a $\phi \in \Upsilon(G)$ such that

$$\|g\| = |\hat{g}(\phi)|.$$

From the sentence immediately after equation 64, we see $\phi \in \text{spt } \mu$. For each $\epsilon > 0$, let

$$U_\epsilon = \{\varsigma \in \mathcal{M}(G): |\hat{g}(\varsigma) - \hat{g}(\phi)| < \epsilon\};$$

clearly U_ϵ is an open set in $\mathcal{M}(G)$. From Propostion 59, we see that there exists an $\epsilon > 0$ such that

$$U_\epsilon \cap \text{spt } \nu_1 = \phi.$$

Using the fact that $\Upsilon(G) \subset \text{spt } \nu$ and that $\text{spt } \nu = \text{spt } \nu_1 \cup \text{spt } \nu_2$, we see that

$$U_\epsilon \cap \Upsilon(G) \subset \text{spt } \nu_2.$$

Now $U_\epsilon \cap \Upsilon(G)$ is a nonempty open subset of $\Upsilon(G)$; it follows then that

$$\hat{\lambda}(U_\epsilon \cap \Upsilon(G)) > 0$$

because $\text{spt } \hat{\lambda} = \Upsilon(G)$. Our contradiction now follows immediately from the next lemma.

■

<u>Lemma</u> 84. Let B be a Borel subset of ∂G such that $\lambda(B) = 0$.

Then

$$\hat{\lambda}(\overline{B^*}) = 0.$$

(In particular, for the proof of the last lemma, we have $\hat{\lambda}(\text{spt}\nu_2) = 0$.)

<u>Proof</u>. Let $\epsilon > 0$ and let $\mathcal{O}$ be an open subset of ∂G such that $\mathcal{O} \supset B$ and $\lambda \mathcal{O} < \epsilon$. Define a function u in $L^\infty(\lambda)$ by

$$u(z) = \begin{cases} 0 & z \in \mathcal{O} \\ 1 & z \in \partial G \setminus \mathcal{O}. \end{cases}$$

Let v be a harmonic conjugate of $\tilde{u}$ in G and let $f = \exp(\tilde{u} + iv)$. Since $C(|f|, z) = \{1\}$ for each z in $\mathcal{O}$, it follows that

$$B^* \subset \overline{B^*} \subset \{\phi \in \mathcal{M}(G): |\hat{f}(\phi)| = 1\}.$$

By the definition of $\hat{\lambda}$, the last set above has $\hat{\lambda}$-measure equal to the $\hat{m}$-measure of

$$A \equiv \{\phi \in \mathcal{M}(D): |(f \circ \psi)\hat{}(\phi)| = 1\}.$$

We claim that $\hat{m}A < \epsilon$. Let us note how this claim will complete the proof of the lemma. Since $\epsilon > 0$ was arbitrary and $\hat{\lambda}(\overline{B^*}) \leq \hat{m}A$, it will follow that $\hat{\lambda}(\overline{B^*}) = 0$.

We shall now prove the claim. Because of the relationship between harmonic measures on ∂D and ∂G, note that

$$(u \circ \psi^*)^{\sim} = \tilde{u} \circ \psi.$$

Now $u \circ \psi^* = \chi_{\psi^{-1}(\partial G \setminus \mathcal{O})} \equiv g$. It is a standard fact that

$$\lim_{r \to 1} \tilde{g}(re^{i\Theta}) = 1$$

for almost every $e^{i\Theta} \in \psi^{-1}(\partial G \setminus \mathcal{O})$. Since

$$m(\psi^{-1}(\partial G \setminus \mathcal{O})) = \lambda(\partial G \setminus \mathcal{O}) > 1 - \epsilon,$$

it follows that $|(f \circ \psi)^*(e^{i\Theta})| = e$ for almost every $e^{i\Theta}$ in a set of measure greater than $1 - \epsilon$. Let

$$C = \{\phi \in \mathcal{M}(D): |(f \circ \psi)\hat{}(\phi)| = e\}$$

and let E be a closed subset of ∂D such that

$(f \circ \psi)^* |_E$ is continuous,

$\mathrm{spt}(m|_E) = E$,

$|(f \circ \psi)^*(e^{i\Theta})| = e$ for every $e^{i\Theta} \in E$,

$m(E) > 1 - \epsilon$.

From Theorem 48 and Theorem 67 we have

$$\mathrm{spt}(m|_{E^*}) \subset C.$$

Thus,

$\hat{m}C \geq \hat{m}(E^*) = m(E) > 1-\epsilon.$

Hence, $\hat{m}(A) < \epsilon$, the claim is established. ■

We have reduced Question 63 for the simply connected case to the following: When does there exist a unital representation π of $H^\infty(G)$ into $L^\infty(\lambda)$ that is an isometry? (We continue our convention of denoting harmonic measure for G at a point z_0 in G by λ. If confusion seems likely, we will add the subscript when the situation calls for it.) Let us now give one answer to the question and also resolve the obvious uniqueness problem.

Theorem 85. If there exists an isometric unital representation of $H^\infty(G)$ into $L^\infty(\lambda)$, then it is unique. Such an isometric representation exists if and only if $\hat{\lambda}$ is a minimal measure in R_λ. When such a representation exsits, then $\hat{\lambda}$ induces the representation.

Before we begin let us note that when G=D, then this theorem is a restatement of Theorem 67. Let us also note that if G is nicely connected, then this theorem shows ι_G is unique and $\hat{\lambda}$ is the minimal measure inducing ι_G.

Proof. Suppose π is an isometric unital representation of $H^\infty(G)$ into $L^\infty(\lambda)$. Let ν be the minimal measure inducing π.

First of all, we claim that spt ν $=\Upsilon(G)$. To see this, repeat the proof of Proposition 74 verbatim (except, of course, one replaces D by G and m by λ) up to the last paragraph. We have to argue differently why the inclusion

$$\text{spt } \nu \subset \overline{(\partial D \setminus K)^*} \cup \bar{U}$$

contradicts the fact spt $\nu \supset \Upsilon(D)$.

To do this, let $u = \chi_K$ and let v be the harmonic conjugate of $\tilde{u}$, the harmonic extension of u to G. Define $f \in H^\infty(G)$ by

$$f = \exp(\tilde{u}+iv).$$

Since $|f| = e^{\tilde{u}}$ and $\lambda(K)>0$, one easily sees that $\|f\| = e$. Since $\partial G \setminus K$ is open and every point on ∂G is regular, it follows that the function h defined on $\overline{G}$ by

$$h = \begin{cases} u & \text{on } \partial G \\ \tilde{u} & \text{on } G \end{cases}$$

is continuous at every point in $\partial G \setminus K$. Using the fact that $C(f,z)=\hat{f}(\mathcal{M}_z)$, one obtains that $|\hat{f}|=1$ on $(\partial D \setminus K)^*$; consequently, $|\hat{f}|=1$ on $\overline{(\partial D \setminus K)^*}$. Since $1<e$, the rest of the proof for the claim is obvious (again, consult the appropriate part of Proposition 74).

Now the spt $\hat{\lambda} = \Upsilon(G)$ and $\hat{\lambda}$ is a normal measure. Appealing to the proof of Proposition 75, we see that $\hat{\lambda}<<\nu$. Since $\hat{\lambda}\in R_\lambda$, we have, by the minimality of ν, that $\hat{\lambda}=\nu$. Therefore, from Theorem 58 we see that π is unique (if it exists). The rest of the details for this theorem now follow in a straight-forward fashion. ∎

Theorem 85 is unsatisfactory in the sense that it may be difficult to see, for a given simply connected domain G, whether $\hat{\lambda}$ is minimal or not. (Without too much difficulty, one can show $\hat{\lambda}$ is not minimal for the slit disc $G \equiv D\setminus\{x\in\mathbb{R}: -1\le x\le 0\}$. Furthermore,the understanding of the lack of minimality for $\hat{\lambda}$ in this canonical case shows what must go wrong when $\hat{\lambda}$ is not minimal for an arbitrary simply connected region.) Our next main objective is to show $\hat{\lambda}$ is minimal if and only if G is nicely connected. Note, if G is nicely connected, then the if statement follows by the comment immediately after Theorem 85.

We now are in position to prove

<u>Lemma</u> 86. Let π be a unital representation of $H^\infty(G)$ into $L^\infty(\lambda)$. If π is an isometry, then π is weak-star, weak-star continuous.

<u>Proof</u>. Suppose π is an isometry. Using essentially the same reasoning as done in the proof of Corollary 16, we see that it suffices to show if $\{f_n\}$ is a sequence of functions in $H^\infty(G)$ that converge weak-star to zero, then $\{\pi(f_n)\}$ converges weak-star to zero in $L^\infty(\lambda)$.

Suppose not; since $\{\|\pi(f_n)\|\}$ is a bounded sequence and the unit ball of $L^\infty(\lambda)$ is weak-star compact (and metrizable), we may assume then, by dropping to a subsequence and relabeling, if need be, that

$$\pi(f_n) \to h \text{ weak-star}$$

where $h \in L^\infty(\lambda)$ and $h \neq 0$. Writing the definition explicitly, we have

$$(87) \quad \int_{\partial G} g\pi(f_n)d\lambda \to \int_{\partial G} ghd\lambda$$

for all $g \in L^1(\lambda)$. As noted in our synopsis of [13, Section 1], for any $z \in G$ there exists a function $g_z \in L^\infty(\lambda)$ such that $\lambda_z = g_z d\lambda$. Combining this fact with Equation (87), we see that

$$\pi(f_n)^\sim(z) \to \tilde{h}(z)$$

for all $z \in G$.

Now fix a $z \in G$. Since $L^\infty(\lambda_z) = L^\infty(\lambda)$, π is an isometry into $L^\infty(\lambda_z)$. From Theorem 85 we see $\hat{\lambda}_z$ induces π. Hence, from equation 47 (with $g=1$) and the results in the paragraph after this equation, we see for any $f \in H^\infty(G)$

$$\int_{\partial G} \pi(f)d\lambda_z = \int_{\mathcal{M}} \hat{f}d\hat{\lambda}_z.$$

Whence, if ψ is the conformal map of D onto G such that $\psi(0)=z$, we have for any $f \in H^\infty(G)$

$$\pi(f)^\sim(z) = \int_{\partial G} \pi(f)d\lambda_z$$

$$= \int_{\mathcal{M}} \hat{f}d\hat{\lambda}_z$$

$$= \int_{\mathcal{M}} (f \circ \psi)^\wedge d\hat{m}$$

(the first $\mathcal{M}$ being $\mathcal{M}(G) \setminus G$, the second being $\mathcal{M}(D) \setminus D$)

$$= f(\psi(0))$$

$= f(z)$.

Since z was arbitrary, we have

$$\pi(f)^{\sim}(z) = f(z)$$

for all $z \in G$ and all $f \in H^{\infty}(G)$.

In particular, we have $\pi(f_n)^{\sim}(z)=f_n(z)$ for all $z\in G$ and all n. Therefore,

$$f_n(z) \to \tilde{h}(z)$$

for all $z\in G$. Since $\{f_n\}$ converges to zero weak-star, it follows that $\tilde{h}(z)=0$ for all $z\in G$. By an easy argument that we leave to the reader, one sees that $h=0$, a contradiction to the fact $h\neq 0$. ∎

<u>We fix some notation now</u>. Let π be an isometric unital representation of $H^{\infty}(G)$ into $L^{\infty}(\lambda)$ where λ is harmonic measure for G at $z_0\in G$. Let ψ be the conformal map of D onto G such that $\psi(0)=z_0$. Hence, ψ^{-1} is a conformal map of G onto D; clearly, $\psi^{-1}\in H^{\infty}(G)$.

<u>Fact</u> 88. $|\pi(\psi^{-1})(z)|=1$ for almost all (with respect to λ) $z\in\partial G$.

<u>Proof</u>. From Theorem 48 we see

$$\pi(\psi^{-1})(z)\in C(\psi^{-1},z)$$

for almost all $z\in\partial G$. Now it is well-known (and easy to prove) that if g is a homeomorphism of G onto D and $\{w_n\}$ is a sequence of points in G that converges to a point on ∂G, then every cluster point of $\{g(w_n)\}$ belongs to ∂D. Combining these two facts together along with the definition of the cluster set $C(\psi^{-1},z)$, we have the fact. ∎

From this fact and Lusin's theorem, we can find an increasing sequence $\{F_n\}$ of compact subsets of ∂G with the following three properties: if $F\equiv \cup F_n$, then $\partial G\setminus F$ has λ-measure zero; for each $z\in F$ one has $|\pi(\psi^{-1})(z)|=1$; and $\pi(\psi^{-1})|_{F_n}$ is continuous for all n. Whenever we refer to $\pi(\psi^{-1})$, we shall take the domain of this function to be

F.

Lemma 89. The function $\pi(\psi^{-1})$ sends sets of positive λ-measure to sets of positive m-measure.

Proof. Suppose not; then there exist an integer n and a compact subset L of F_n such that $\lambda(L)>0$ and $m(\pi(\psi^{-1})(L))=0$. We define a measure β on ∂D with spt $\beta \subset \pi(\psi^{-1})(L)$ as follows: For all Borel sets $A \subset \pi(\psi^{-1})(L)$, let

$$\beta(A) = \lambda(\pi(\psi^{-1})^{-1}(A)).$$

Clearly, $\beta \neq 0$ and $\beta \perp m$.

Using [37] and a standard argument, we can find a sequence $\{p_n\}$ of polynomials such that p_n converges to ψ^* weak-star in $H^\infty(m)$ and p_n converges to zero pointwise almost everywhere β. Note then that p_n converges to ψ weak-star in $H^\infty(D)$. Whence, $p_n \circ \psi^{-1}$ converges to χ weak-star in $H^\infty(G)$. By Lemma 86 we see then $\pi(p_n \circ \psi^{-1})$ converges weak-star to χ in $L^\infty(\lambda)$. In particular, we see

(90) $\pi(p_n \circ \psi^{-1})\,|_L \to \chi\,|_L$

weak-star in $L^\infty(\lambda\,|_L)$.

On the other hand, if we note $\pi(p_n \circ \psi^{-1}) = p_n(\pi(\psi^{-1}))$ for all p_n (because they are polynomials), then

$$\pi(p_n \circ \psi^{-1}) \to 0$$

pointwise almost everywhere $\lambda\,|_L$ because $p_n \to 0$ pointwise almost everywhere β. Using the fact $\{\|p_n\|\}$ is a bounded sequence, we then have, by dominated convergence,

$$\pi(p_n \circ \psi^{-1}) \to 0$$

weak-star in $L^\infty(\lambda\,|_L)$. This contradicts the result obtained in (90) because $\lambda(L)>0$.

■

Lemma 91. We have

$$m(\partial D \setminus \pi(\psi^{-1})(F)) = 0.$$

Proof. Suppose not; then there exists a compact set $A \subset \partial D \setminus \pi(\psi^{-1})(F)$ such that $m(A)>0$. Choose a function $f \in H^{\infty}(D)$ such that

$$|f^*| = \begin{cases} 2 & \text{on } A \\ 1 & \text{on } \partial D \setminus A \end{cases}$$

(equality meaning m almost everywhere). We may further assume that the function

$$h = \begin{cases} f_* & \text{on } D \\ f^* & \text{on } \partial D \end{cases}$$

is continuous on $\partial D \setminus A$. Note, since π is an isometry, that $\|\pi(f \circ \psi^{-1})\| = \|f \circ \psi^{-1}\| = 2$. From Lemma 89 we see $f^* \circ \pi(\psi^{-1})$ is a λ-measurable function. By the definition of A (a subset of $\partial D \setminus \pi(\psi^{-1})(F)$) we clearly see that $\|f^* \circ \pi(\psi^{-1})\| = 1$. The desired contradiction is close at hand: choose a sequence of polynomials $\{p_n\}$ such that the sequence is bounded and converges pointwise m-almost everywhere on ∂D to f^*. Hence, $p_n \to f$ weak-star in $H^{\infty}(D)$. Therefore, by Lemma 86, we have

$$\pi(p_n \circ \psi^{-1}) \to \pi(f \circ \psi^{-1})$$

weak-star in $L^{\infty}(\lambda)$. But,

$$p_n(\pi(\psi^{-1})) \to f^*(\pi(\psi^{-1}))$$

pointwise λ-almost everywhere. Hence, using the fact $\pi(p_n \circ \psi^{-1}) = p_n(\pi(\psi^{-1}))$ for all n, we see

$$(92) \quad \pi(f \circ \psi^{-1}) = f^*(\pi(\psi^{-1}))$$

λ almost everywhere. Taking the norms of both sides of this last equality, we see $2=1$; an absurdity. ∎

Note that the argument used in establishing (92) did not use the property of the particular $f \in H^{\infty}(D)$ constructed in the proof. We actually established that (if π is an isometry)

$$\pi(f\circ\psi^{-1}) = f^*(\pi(\psi^{-1}))$$

for all $f\in H^\infty(D)$. Using $f=\psi$ in this last equality and applying the result of the last lemma, we have the following.

Theorem 93. If π is a unital representation of $H^\infty(G)$ into $L^\infty(\lambda)$, then G is nicely connected.

The following theorem is a summary of the results obtained.

Theorem 94. Let G be a bounded simply connected region. The region G supports an isometric unital representation if and only if G is nicely connected. If G is nicely connected, then the minimal measure inducing the (natural) isometric unital representation ι_G of $H^\infty(G)$ into $L^\infty(\lambda)$ is $\hat{\lambda}$. Furthermore, if G is nicely connected, then π, a unital representation of $H^\infty(G)$ into $L^\infty(\mu)$, is an isometry if and only if each of the following conditions hold:

(x) If

$$\mu = \mu_1 + \mu_2$$

is the Lebesgue decomposition of μ with respect to λ, then μ_1 and λ are equivalent and $\mu_2\perp\lambda$. Hence, $L^\infty(\mu)=L^\infty(\lambda)\oplus L^\infty(\mu_2)$.

(xi) $\pi = \iota_G\oplus\pi_2$

where π_2 is some unital representation of $H^\infty(G)$ into $L^\infty(\mu_2)$.

CHAPTER VII

PARTIALLY SUBORDINATE REPRESENTATIONS

In this section we present a few results related to the earlier work of the paper. To begin, recall if A is any set of operators in $\mathcal{B}(\mathcal{H})$, then the lattice of A, denoted Lat A, is the collection of closed subspaces $\mathcal{L}$ of $\mathcal{H}$ such that $T\mathcal{L}\subset\mathcal{L}$ for all $T\in A$. If $\mathcal{A}$ is the weakly closed algebra generated by A and 1, then

$\mathrm{Lat}\ A = \mathrm{Lat}\ \mathcal{A}$

Fix a bounded region G and a unital representation π from $H^{\infty}(G)$ into $L^{\infty}(\mu)$ where μ is a probability measure with spt $\mu \subset \partial G$. As noted in the introduction, we can associate with π another representation $\tilde{\pi}$ from $H^{\infty}(G)$ into $\mathcal{B}(L^2(\mu))$ in the obvious way:

$$\tilde{\pi}(f) \equiv M_{\pi(f)} \text{ on } L^2(\mu)$$

for all $f \in H^{\infty}(G)$. If

$$\mathcal{A}_{\tilde{\pi}} \equiv \mathrm{ran}\ \tilde{\pi},$$

then what is $\mathrm{Lat}\mathcal{A}_{\tilde{\pi}}$?

If E is a Borel subset of ∂G, then clearly $\chi_E L^2(\mu)$ belongs to $\mathrm{Lat}\mathcal{A}_{\tilde{\pi}}$. Recall χ_E denotes the characteristic function of E (which should not be confused with the function $\chi|_E$, the restriction of χ to E). What other subspaces belong to $\mathrm{Lat}\ \mathcal{A}_{\tilde{\pi}}$?

Suppose $\mathcal{L}$ is a closed subspace of $L^2(\mu)$, $M_{\chi}\mathcal{L} \subset \mathcal{L}$ and $M_{\chi}|_{\mathcal{L}}$ is a normal operator. It is well-known (and easy to prove) that $\mathcal{L}$ then is reduced by M_{χ} (definition: invariant under M_{χ} and $M_{\chi}^{*} = M_{\bar{\chi}}$). Hence, the projection of $\mathcal{H}$ onto $\mathcal{L}$ commutes with M_{χ}. Since $\{M_g\colon g \in L^{\infty}(\mu)\}$ is a maximal abelian subalgebra of $\mathcal{B}(L^2(\mu))$, one can easily see that $\mathcal{L} = \chi_E\ L^2(\mu)$ for some Borel subset E of ∂G. Note then that $\mathcal{L}$ is a reducing subspace for M_g, for all $g \in L^{\infty}(\mu)$

Now suppose $\mathcal{L} \in \mathrm{Lat}\ \mathcal{A}_{\tilde{\pi}}$. In particular, $M_{\chi}\mathcal{L} \subset \mathcal{L}$; hence, $S = M_{\chi}|_{\mathcal{L}}$ is a subnormal operator. It is a routine exercise to show that there exist two unique orthogonal subspaces $\mathcal{L}_i \in \mathrm{Lat}\ S$ such that

$$\mathcal{L} = \mathcal{L}_1 \oplus \mathcal{L}_2,$$

$S|_{\mathcal{L}_1}$ is pure and $S|_{\mathcal{L}_2}$ is normal. (One allows the possibility that one of the $\mathcal{L}_i$ may be $\mathcal{L}$ itself.) From the preceding paragraph we see that $\mathcal{L}_i \in \mathrm{Lat}\ \mathcal{A}_{\tilde{\pi}}$ for $i=1,2$.

We then draw the conclusion that

$$\text{Lat } \mathcal{A}_{\tilde{\pi}} \underset{\neq}{\supset} \{\chi_E L^2(\mu)\text{: E a Borel subset of } \partial G\}$$

if and only if there exists $\mathcal{L} \in \text{Lat } \mathcal{A}_{\tilde{\pi}}$ such that $S = M_\chi|_{\mathcal{L}}$ is a pure subnormal operator. Let us suppose this is the case; let N be the minimal normal extension of S. That is, $N = M_\chi$ restricted to the (reducing) subspace

$$\mathcal{K} = \{\sum_{j=0}^{n} (M_\chi)^{*j} x_j\text{: } x_j \in \mathcal{L}, n \geq 0\}^{-L^2(\mu)}.$$

From the above we see there exists a Borel set $A \subset \partial G$ such that

$$\mathcal{K} = \chi_A L^2(\mu);$$

so $N = M_\chi$ on $L^2(\mu|_A)$. Clearly, $\mathcal{K} \in \text{Lat}\mathcal{A}_{\tilde{\pi}}$. (In fact, K reduced $\tilde{\pi}(f)$ for all $f \in H^\infty(G)$.) Hence, the mapping

$$f \to \tilde{\pi}(f)|_{\mathcal{K}} \text{ for } f \in H^\infty(G)$$

is a unital representation of $H^\infty(G)$ into $\mathcal{B}(L^2(\mu|_A))$ where each $\tilde{\pi}(f)|_{\mathcal{L}}$ is a normal operator.

We now want to show that the representation π_1 from $H^\infty(G)$ into $L^\infty(\mu|_A)$ defined by

$$\pi_1(f) \equiv \pi(f)|_A$$

is an isometry and is weak-star, weak-star continuous. To do this, we first make some elementary observations. Note that the representation

$$f \to \tilde{\pi}(f)|_{\mathcal{K}}$$

is, from the definition of ~, the representation $\tilde{\pi}_1$. That is, for each $f \in H^\infty(G)$,

(95) $\tilde{\pi}_1(f) = M_{\pi_1(f)}$ on $L^2(\mu|_A)$.

It follows then, for every $f \in H^\infty(G)$,

$$\tilde{\pi}_1(f)\mathcal{L} \subset \mathcal{L}$$

and

$$\tilde{\pi}_1(f)|_{\mathcal{L}} = \tilde{\pi}(f)|_{\mathcal{L}}.$$

Let η be the obvious unital representation of $H^\infty(G)$ into $\mathcal{B}(\mathcal{L})$ such that $\eta(\chi)=S$ (which is a pure subnormal operator): for each $f \in H^\infty(G)$, define

$$\eta(f) \equiv \tilde{\pi}_1(f)|_{\mathcal{L}}.$$

To see that π_1 is an isometry and is weak-star, weak-star continuous, one must show $\tilde{\pi}_1$ has these two properties (consult (95)). To do the latter, using [10, Corollary 2.16, p. 134] and [10, Theorem 12.9, p. 207] in conjunction with the definition of η and the fact $\tilde{\pi}_1(\chi)$ is the minimal normal extension of $\eta(\chi)$, one sees it suffices to show η is an isometry and is weak-star, weak-star continuous.

The weak-star continuity of η follows immediately from Corollary 15 because $\eta(\chi)=S$ is pure. To see that η is an isometry, it is enough to show that $\sigma(S)=\bar{G}$. (The fact then follows from Corollary 22 and the result that the norm of a subnormal operator is its spectral radius.) Since η is a unital representation of $H^\infty(G)$ into $\mathcal{B}(\mathcal{L})$ with $\eta(\chi)=S$, clearly $\sigma(S) \subset \bar{G}$. Now $\tilde{\pi}_1(\chi)$ acting on $L^2(\mu|_A)$ is the minimal normal extension of S. From the Bram–Halmos theorem [10, Theorem 2.11, p. 131], it now follows that either $\sigma(S) \subset \partial G$, or $\sigma(S)=\bar{G}$. The first possibility cannot occur; otherwise, from Lemma 9 and [10, Theorem 1.1, p. 302] the fact that S is normal would arise. We have established that π_1 has the properties claimed.

Let us introduce some notation that will be helpful to this discussion. Let π_i, $i=1,2$, be two unital representations of $H^\infty(G)$ into $L^\infty(\mu_i)$ where spt $\mu_i \subset \partial G$. _We say_ π_2 _is subordinate to_ π_1 if μ_2 is absolutely continuous with respect to μ_1 and

$$\pi_2(f) = \pi_1(f)$$

almost everywhere μ_2, for all $f \in H^\infty(G)$. It is easy to see that π_2 is subordinate to π_1

if there exist measures μ_3 and μ_4 with $\mu_3 \perp \mu_2$ and μ_4 and μ_2 mutually absolutely continuous such that

$$\mu_1 = \mu_4 + \mu_3,$$

and there exists a unital representation π_3 from $H^\infty(G)$ into $L^\infty(\mu_3)$ such that

$$\pi_1 = \pi_4 \oplus \pi_3 = \pi_2 \oplus \pi_3.$$

In the discussion above we showed that π_1 was subordinate to π provided

$$\text{Lat } \mathcal{A}_{\tilde{\pi}} \neq \{\chi_E L^2(\mu) : E \text{ a Borel subset of } \partial G\}.$$

<u>Theorem</u> 96. Let G be a bounded region and let π be a unital representation of $H^\infty(G)$ into $L^\infty(\mu)$ where spt $\mu \subset \partial G$. Let $\tilde{\pi}$ be the (natural) associated representation of $H^\infty(G)$ into $\mathcal{B}(L^2(\mu))$. The following conditions are equivalent:

(i) Lat $\mathcal{A}_{\tilde{\pi}} \neq \{\chi_E L^2(\mu)$: E a Borel subset of $\partial G\}$.

(ii) There exists $\mathcal{L} \in \text{Lat } \mathcal{A}_{\tilde{\pi}}$ such that $M_\chi |_{\mathcal{L}}$ is a pure subnormal operator.

(iii) There exists an isometric unital representation π_1 of $H^\infty(G)$ into $L^\infty(\nu)$ that is weak-star, weak-star continuous and π_1 is subordinate to π.

(iv) Ran π is not weak-star dense in $L^\infty(\mu)$.

<u>Proof</u>. We have already established the equivalence of (i) and (ii) and the fact that (ii) implies (iii). Let us now show (iii) implies (iv).

Without loss of generality, we may assume there exists a Borel subset $A \subset \partial G$ such that $\nu = \mu|_A$, and there exists a unital representation π_2 of $H^\infty(G)$ into $L^\infty(\mu|_{\partial G \setminus A})$ such that

$$\pi = \pi_1 \oplus \pi_2.$$

We claim that ran π_1 is a proper subalgebra of $L^\infty(\mu|_A)$. To see this, first observe the following elementary result (whose proof is left to the reader). Let β be a positive regular Borel measure with compact support. The multiplicative linear functionals on

$L^\infty(\beta)$ that are weak-star continuous arise precisely as evaluations at all points x in the support of β where $\beta(\{x\})>0$. Now fix $z_0 \in G$. The mapping

$$f \to f(z_0)$$

for $f\in H^\infty(G)$ is a weak-star continuous multiplicative linear functional on $H^\infty(G)$. Since ran π_1 is weak-star homeomorphic to $H^\infty(G)$ under π_1, the mapping

$$\pi_1(f) \to f(z_0)$$

for $f\in H^\infty(G)$ is a weak-star continuous multiplicative linear functional on ran π_1. Noting the elementary result, we see the validity of the claim.

Observing that ran π_1 is weak-star closed (because $H^\infty(G)$ is weak-star closed), it follows that

$$(\text{ran } \pi)^{-wk^*} \subset (\text{ran } \pi_1) \oplus (L^\infty(\mu \mid_{\partial G \setminus A});$$

whence, from the claim the assertion in (iv) follows.

To see that (iv) implies (i), one only needs to apply the fact that any weakly closed algebra of normal operators which contains 1 is reflexive [35, Theorem 9.21] in conjunction with the fact the set on the right-hand side of the inequality in (i) is precisely

$$\text{Lat}\{M_g \colon g\in L^\infty(\mu)\}. \quad ■$$

Combining the results in Theorems 96 and 94, we obtain

Corollary 97. Let G be a bounded simply connected region. Let π be a unital representation of $H^\infty(G)$ into $L^\infty(\mu)$ where spt $\mu\subset\partial G$. The following conditions are equivalent:

(v) π is an isometry.

(vi) Ran π is not weak-star dense in $L^\infty(\mu)$.

(vii) The region G is nicely connected and the natural representation ι_G of $H^\infty(G)$ into $L^\infty(\lambda)$ is a subordinate to π.

(viii) Lat $\mathcal{A}_{\tilde{\pi}} \neq \{\chi_E L^2(\mu)\colon$ E a Borel subset of $\partial G\}$.

(ix) There exists $\mathcal{L}\in\mathrm{Lat}\ \mathcal{A}_{\tilde{\pi}}$ such that $M_\chi|_{\mathcal{L}}$ is a pure subnormal operator.

The authors believe that the equivalence of (v) and (vi) holds for an arbitrary domain. (Note, (vi) implies (v) follows from Theorem 96; reason, any representation π, that has an isometric representation subordinate to π, must be an isometry. Perhaps, there exists a result from the general theorey of Banach algebras (which the authors are unaware of) that shows the other direction ((v)=>(vi)) is elementary too.) We note that verification of the implication (vi)=>(v) relied heavily on some tools from operator theory. However, these conditions are purely function-theoretic. A proof (like that done in the verification of Corollary 66) based on the concept of minimal measure would be interesting.

A simple consequence of Corollary 97 is the following result of Chevreau-Pearcy-Shields [8].

Corollary 98. Any algebra endomorphism of $H^\infty(D)$ that leaves each polynomial fixed is the identity.

Proof. Clearly the statement to be proved is equivalent to the following one:

(99) If π is a unital representation of $H^\infty(D)$ into $H^\infty(m)$, then $\pi=\iota_D$.

To see the validity of the latter statement, note that ran π is clearly not weak-star dense in $L^\infty(m)$. Hence, from the previous corollary π is an isometry. The result now follows from Theorem 67. Let us remark that the validity of (99) does not require any extensive theory. Let us give another (elementary) proof of (99).

For $g\in L^\infty(m)$ the harmonic extension of g to D will be denoted by $\tilde{g}$. To verify (99), clearly it suffices to show, for each $z\in D$,

$$f(z) = (\pi f)^{\sim}(z)$$

for all $f\in H^\infty(D)$. Fix $z\in D$. The mapping

(100) $f \rightarrow (\pi f)^{\sim}(z)$

is a multiplicative linear functional on $H^\infty(D)$ because π is multiplicative and $\sim$ is multiplicative on $H^\infty(m)$. Now

$$\chi \rightarrow \tilde{\chi}(z) = z,$$

so the multiplicative linear functional on $H^{\infty}(D)$ given in (100) is evaluation at z. (There is only one element in the fiber of $\mathcal{M}(D)$ above a point $z \in D$.) That is, for all $f \in H^{\infty}(D)$

$$f(z) = (\pi f)^{\sim}(z). \blacksquare$$

Let us return to the theme of the beginning of this section. If G is simply connected and π is a unital representation of $H^{\infty}(G)$ into $L^{\infty}(\mu)$, we can give an answer to the basic problem of computing Lat $\mathcal{A}_{\tilde{\pi}}$. If G is not nicely connected, then

$$\text{Lat } \mathcal{A}_{\tilde{\pi}} = \{\chi_E L^2(\mu): \text{E a Borel subset of } \partial G\}.$$

If G is nicely connected and ι_G is not subordinate to π, the answer remains the same. Without any harm, we may now assume

(x) $\mu = \lambda + \nu$ where $\nu \perp \lambda$, and

(xi) $\pi = \iota_G \oplus \pi_1$ where π_1 is a unital representation of $H^{\infty}(G)$ into $L^{\infty}(\nu)$.

First of all, we note

(101) $\text{Lat } \mathcal{A}_{\tilde{\pi}_1} = \{\chi_E L^2(\nu): \text{E a Borel subset of } \partial G\}.$

Justification: From Theorem 94 we see that no isometry is subordinate to π_1; whence, from Corollary 97 the note follows.

We now claim that

(102) $\text{Lat } \mathcal{A}_{\tilde{\pi}} = \text{Lat } \mathcal{A}_{\iota_G} \oplus \text{Lat } \mathcal{A}_{\tilde{\pi}_1};$

i.e., the lattice of $\mathcal{A}_{\tilde{\pi}}$ splits according to the decomposition of (xi).

To verify (102), let F be a σ-compact subset of ∂G such that ν is carried by F, and let t denote the characteristic function of $\partial G \setminus F$. We shall show that $M_t \in \mathcal{A}_{\tilde{\pi}}^{-wot}$. Using the fact that any weakly closed algebra of normal operators that contains 1 is reflexive, the reader can then show (easily) how (102) follows.

Let A(G) denote the uniform algebra of continuous functions on $\overline{G}$ that are analytic on G. From [13], we see that A(G) is a Dirichlet algebra on ∂G. It follows then that λ is a unique representing measure for A(G) at z_0 (where λ is harmonic measure for $z_0 \in G$). Therefore, from [1] there exists a sequence $\{f_n\} \subset A(G)$ such that $\|f_n\| \le 1$ for all n,

$$f_n \to 0$$

pointwise everywhere on F, and

$$f_n \to 1$$

pointwise almost everywhere with respect to λ. It follows easily from Theorem 48 that $\pi(f_n) = f_n$ for all n. Hence,

$$\pi(f_n) \to t$$

weak-star in $L^\infty(\mu)$ by the dominated convergence theorem. Whence, $M_t \in \mathcal{A}_{\tilde{\pi}}^{-wot}$.

Combining (101) and (102), we have established the following.

<u>Theorem</u> 103. Let G be a simply connected domain. Let π be a unital representation of $H^\infty(G)$ into $L^\infty(\mu)$. Then

$$\text{Lat } \mathcal{A}_{\tilde{\pi}} = \{\chi_E L^2(\mu)\colon \text{E a Borel subset of } \partial G\}$$

unless G is nicely connected and ι_G is subordinate to π. In this case, if

$$\mu = \mu_1 + \mu_2$$

is the Lebesgue decomposition of μ with respect to λ where μ_1 and λ are equivalent measures and $\mu_2 \perp \lambda$, then

$$\text{Lat } \mathcal{A}_{\tilde{\pi}} = \text{Lat } \mathcal{A}_{\iota_G} \oplus \{\chi_E L^2(\mu_2)\colon \text{E a Borel subset of } \partial G\}.$$

A few words seem in order concerning the structure of Lat $\mathcal{A}_{\iota_G}$. Let

$$\mathcal{H} = \{\iota_G(f)\colon f \in H^\infty(G)\}^{-L^2(\lambda)}$$

and let S be the subnormal operator

$$S \equiv M_\chi|_{\mathcal{H}}.$$

One sees that S is a pure subnormal operator and

$$\text{Lat } S = \text{Lat } \mathcal{A}_{\tilde{\iota}_G}.$$

Let ψ be the conformal map of D onto G such that $\psi(0)=z_0$ where λ is harmonic measure at z_0. From our synopsis of [13] given in Section 6, since G is nicely connected, it follows that S is unitarily equivalent to the analytic Toeplitx operator on $H^\infty(m)$ whose symbol is ψ. Let T_ψ denote this latter operator. If $\mathcal{L}_1$ and $\mathcal{L}_2$ are two isomorphic lattices, then we will write $\mathcal{L}_1 \cong \mathcal{L}_2$. We have then the following result.

Proposition 104. With the assumptions and notation introduced in the preceding paragraph, we have

$$\text{Lat } \mathcal{A}_{\tilde{\iota}_G} \cong \text{Lat } T_\psi.$$

Our second topic of this section deals with the subject matter of Sections 5 and 6. Suppose G is a bounded domain and π is a unital representation of $H^\infty(G)$ into $L^\infty(\mu)$ where μ is a probability measure with spt $\mu \subset \partial G$. How can you tell when π is a one-to one-mapping? Is there a condition on ν, the minimal measure inducing π, that characterizes when π is one-to-one? We have not found the answers to these questions. We finish the paper by restricting our attention to this matter when π is a unital representation defined on $H^\infty(D)$. The reader may look at Example 77 and predict for himself (herself) what the outcome will be.

Before we do this, we need to develop a little more theory about unital representations. Let π_i, for i=1,2, be a unital representation of $H^\infty(G)$ into $L^\infty(\mu_i)$. We say π_1 and π_2 are partially subordinate if the measures μ_i are not mutually singular and there exists a Borel subset $A \subset \partial G$ with $d\mu_1/d\mu_2 > 0$ everywhere on A, $\mu_2(A) > 0$ and

$$\pi_2(f) = \pi_1(f)$$

almost everywhere with respect to $\mu_2|_A$, for all $f \in H^\infty(G)$. Loosely speaking, the representations π_i are partially subordinate if somewhere locally the representations are equal. We now want to characterize partially subordinate representations in terms of their minimal measures. The characterization depends on the next proposition whose proof is left as an exercise for the reader.

Proposition 105. Let G be a bounded region and let μ be a measure with spt $\mu\subset\partial G$. If ν is a minimal measure in R_μ and E is a Borel subset of ∂G with $\mu(E)>0$, then $\nu|_{E^*}$ is a minimal measure in $R_{\mu|_E}$. (The notation used is explained in Section 5.)

We now are ready to prove the following.

Theorem 106. Let G be a bounded region and let π_i, for i=1,2, be a unital representation of $H^\infty(G)$ into $L^\infty(\mu_i)$. Let ν_i be the minimal measure inducing π_i for i=1,2. The representations π_1 and π_2 are partially subordinate if and only if ν_1 and ν_2 are not mutually singular. The representation π_1 is subordinate to π_2 if and only if ν_1 is absolutely continuous with respect to ν_2.

Proof. We shall prove the characterization of partial subordination; the details of the other part of the theorem follow by a similar argument.

Suppose π_1 and π_2 are partially subordinate and let A be a Borel subset of ∂G such that $\mu_1|_A$ and $\mu_2|_A$ are equivalent measures (mutually absolutely continuous), $\mu_1(A)>0$, and

(107) $\pi_1(f) = \pi_2(f)$

almost everywhere $\mu_1|_A$, for all $f\in H^\infty(G)$. Clearly then $\nu_1|_{A^*}$ and $\nu_2|_{A^*}$ are equivalent measures. It now follows in a straight-forward fashion from this last fact, equation (107) and Theorem 58 that $\nu_1|_{A^*}$ and $\nu_2|_{A^*}$ are mutually absolutely continuous (hence, ν_1 and ν_2 are not singular).

Now suppose that ν_1 and ν_2 are not singular measures. Take the Lebesgue decomposition of ν_2 with respect to ν_1; say

$$\nu_2 = h\nu_1 + \nu_s$$

where $h\in L^1(\nu_1)$, $h\neq 0$ and $\nu_s\perp\nu_1$. Let T be a closed subset of $\mathcal{M}(=\mathcal{M}(G)\setminus G)$ such that $h>0$ almost everywhere $\nu_1|_T$, $\nu_1(T)>0$ and $\nu_s(T)=0$. Let γ be the measure defined on ∂G by

$$\gamma(E) = \nu_1(E^* \cap T)$$

for all Borel subsets $E \subset \partial G$. Since

$$\nu_2|_T = h\nu_1|_T$$

and $h>0$ almost everywhere $\nu_1|_T$, it follows that $\nu_2|_T$ and $\nu_1|_T$ are equivalent measures. By Lemma 49 there exists a measurable subset $A \subset \partial G$ such that

$$\gamma = \chi_A \mu_1.$$

Let E be a compact subset of A with $\mu_1(E)>0$. It follows now that

$$\mathrm{spt}(\nu_1|_{F^*}) \subset T$$

for all closed subsets F of E. Hence, since $\nu_1|_T$ and $\nu_2|_T$ are equivalent, we see that $\mu_1|_E$ and $\mu_2|_E$ are equivalent measures.

Using Proposition 105, we see that $\nu_1|_{E^*}$ and $\nu_2|_{E^*}$ are minimal measures with respect to $\mu_1|_E$ and $\mu_2|_E$, respectively. Hence, they induce unital representations π_3 and π_4 from $H^\infty(G)$ into $L^\infty(\mu_1|_E)$ and $L^\infty(\mu_2|_E)$, respectively. Note, $L^\infty(\mu_1|_E)$ and $L^\infty(\mu_2|_E)$ are equal spaces because of the equivalence of $\mu_1|_E$ and $\mu_2|_E$. Now, since $\nu_1|_T$ and $\nu_2|_T$ are also equivalent and

$$\mathrm{spt}(\nu_1|_{E^*}) \subset T,$$

it follows that

$$\mathrm{spt}(\nu_1|_{F^*}) \subset \mathrm{spt}(\nu_2|_{F^*})$$

for all closed subsets F of E. Since $\nu_2|_{E^*}$ is minimal for $\mu_2|_E$, one sees the last inclusion is actually equality. Thus, $\pi_3=\pi_4$; whence, π_1 and π_2 are partially subordinate.

■

We now direct our efforts in determining which unital representations π defined on

$H^\infty(D)$ are one-to-one. Our reward will be the following theorem.

Theorem 108. Let π be a unital representation of $H^\infty(D)$ into $L^\infty(\mu)$ where spt $\mu \subset \partial D$. The representation π is one-to-one if and only if π and ι_D are partially subordinate. Equivalently, π is one-to-one if and only if there exists a Borel subset $E \subset \partial D$ with $m(E>0$, $\mu|_E$ and $m|_E$ are equivalent measures, and

$$\pi(f)|_E = \iota_D(f)|_E = f^*|_E$$

almost everywhere m, for all $f \in H^\infty(D)$. (Recall, f^* is that function in $H^\infty(m)$ obtained by taking radial limits of f.)

We begin with the following result of D.J. Newman [29]. We do not prove the fact; however, we have used his technique in some other proofs to follow.

Lemma (D.J. Newman) 109. Let B be a Blaschke product. There exists a Blaschke product q such that whenever $\psi \in \mathcal{M} = \mathcal{M}(D) \setminus D$ and

$$|\psi(B)| < 1,$$

then

$$\psi(q) = 0.$$

From now on, $\mathcal{M} = \mathcal{M}(D) \setminus D$ and $\Upsilon = \Upsilon(D)$.

Lemma 110. Let T be a compact subset of $\mathcal{M} \setminus \Upsilon$. There exists a Blaschke product B such that

$$\phi(B) = 0 \text{ for all } \phi \in T.$$

Proof. Let $\psi \in T$. By [28, Theorem on p. 179], there exists a Blaschke product B_ψ such that $\psi(B_\psi)=0$. Let

$$N(\psi) = \{\phi \in \mathcal{M}: |\phi(B_\psi)| < 1\}.$$

Since T is compact and $N(\psi)$ is open for all $\psi \in T$, there exists a finite subset $\{\psi_1, \ldots, \psi_n\}$ of T such that

$$T \subset \bigcup_{i=1}^{n} N(\psi_i).$$

From Lemma 109, there exists, for each i, a Blaschke product B_i such that

$$\phi(B_i) = 0$$

for all $\phi \in N(\psi_i)$. Let

$$B = \prod_{i=1}^{n} B_i.$$

Clearly, $\phi(B)=0$ for all $\phi \in T$. ∎

We now improve the last lemma to σ-compact subsets of $\mathcal{M} \setminus \Upsilon$.

Lemma 111. Let T be a σ-compact subset of $\mathcal{M} \setminus \Upsilon$. There exists a Blaschke product B such that

$$\phi(B) = 0$$

for all $\phi \in T$.

Proof. We write

$$T = \bigcup_n T_n$$

where $\{T_n\}$ is a sequence of compact subsets of T. Fix an n. From the last lemma, there exists a Blaschke product B_n such that $\hat{B}_n|_{T_n} = 0$. Clearly B_n must be an infinite Blaschke product, so we can write

$$B_n = \prod_{k=1}^{\infty} \frac{\bar{\alpha}_k}{|\alpha_k|} \frac{\alpha_k - \chi}{1 - \bar{\alpha}_k \chi},$$

where $\{\alpha_k\}$ is the sequence of zeros of B_n, listed according to their multiplicities. Let

$$g_{nj} = \prod_{k=j}^{\infty} \frac{\bar{\alpha}_k}{|\alpha_k|} \frac{\alpha_k - \chi}{1 - \bar{\alpha}_k \chi}.$$

It is easy to see

$$|\phi(g_{nj})| = |\phi(B_n)|$$

for all $\phi \in \mathcal{M}$. Choose a j so large that

$$\sum_{k=j}^{\infty} (1 - |\alpha_k|) < (1/2^n).$$

Let $p_n = g_{nj}$.

Now let B be the Blaschke product whose zero set is the union of the zero sets of the p_n's, counting multiplicities. B is the desired Blaschke product; i.e.,

$$\hat{B}|_T = 0.$$

To see this, note that for each n there exists a Blaschke product r_n such that

$$B = p_n r_n.$$

Thus, for all $\phi \in T_n$,

$$\phi(B) = \phi(p_n)\phi(r_n) = 0. \blacksquare$$

We now establish an analogous result for a σ-compact $T \subset \Upsilon$ with $\hat{m}(T)=0$. Recall that $\hat{m}$ is the minimal measure inducing ι_D.

Lemma 112. Let T be a compact subset of Υ such $\hat{m}(T)=0$. There exists a nonzero function $f \in H^\infty(D)$ with $\|f\| \leq 1$ and

$$\hat{f}|_T = 0.$$

Proof. For positive integers k and n define a function s_{nk} in $C(\Upsilon)$ satisfying

(xii) $-n \leq s_{nk} \leq 0.$

(xiii) $s_{nk} = -n$ on T.

(xiv) $\hat{m}\{\phi: s_{nk}(\phi)=0\} > 1-k^{-1}.$

Let u_{nk} be that function in $L^\infty(m)$ such that $\hat{u}_{nk}$, the Gelfand transform of u_{nk}, is s_{nk}. (As before, we identify the maximal ideal space of $L^\infty(m)$ with Υ.) Choose a sequence $\{a_n\}$ of positive numbers such that Σa_n converges. Fix an n. We observe that

$$\tilde{u}_{nk}(0) = \int u_{nk} dm$$

$$= \int s_{nk} d\hat{m} \to 0$$

as $k \to \infty$. Choose k so large that

$$\tilde{u}_{nk}(0) \geq -a_n.$$

Let $u_n = u_{nk}$ for that fixed k; we do this for all n.

Let v_n be the harmonic conjugate of $\widetilde{u}_n$ and let

$$g_n = \exp(\widetilde{u}_n + iv_n).$$

Clearly, $g_n \in H^\infty(D)$ and $\|g_n\| \leq 1$. Let

$$f_i = \prod_{n=1}^{i} g_n.$$

Now $\|f_i\| \leq 1$ for all i, so $\{f_i\}$ is a normal family. Choose a subsequence $\{f_{i_j}\}$ that converges uniformly on compacta to an H^∞-function f. Since

$$|g_n(0)| = \exp(\widetilde{u}_n(0)) > \exp(-a_n),$$

it follows that

$$|f_i(0)| \geq \exp(-a_1 - a_2 - \dots - a_i)$$

$$\geq \exp(-\Sigma a_n).$$

Thus, $|f(0)| > 0$ and $\|f\| \leq 1$.

It only remains to show

$$\hat{f}|_{\Upsilon} = 0.$$

It is easy to see that, for all i,

$$(f/f_i) \in H^\infty(D),\ \|f/f_i\| \leq 1, \text{ and } f = (f/f_i)f_i.$$

Taking the Gelfand transform of both sides of the last inequality, we see that

$$|\hat{f}| \leq |\hat{f}_i|$$

everywhere on $\mathcal{M}$. Note that on ∂D we have

$$|g_n^*| = e^{u_n}$$

almost everywhere. Using the fact that Gelfand transform is an isometric *-algebra homomorphism on $L^\infty(m)$, we see that on Υ

$$|\hat{g}_n| = |g_n^*|^\wedge$$

$$= (\exp(u_n))^\wedge$$

$= \exp(\hat{u}_n)$.

Now on T we have $\hat{u}_n = -n$; hence, on T we have

$$|\hat{g}_n| = \exp(-n).$$

Thus, on T for all i we have

$$|\hat{f}| \le |\hat{f}_i|$$
$$\le \exp(-1-2-\ldots-i).$$

It follows that $\hat{f}|_T = 0$. ∎

Remark 113. The converse of the last lemma is true; i.e., if T is a compact subset of Υ and there exists $f \in H^\infty(D)$ with $f \ne 0$ such that

$$\hat{f}|_T = 0,$$

then $\hat{m}(T)=0$. (One way to see this is to observe that for any nonzero $f \in H^\infty(D)$ the measure $|\hat{f}|d\hat{m}$ is the minimal measure inducing the representation ι_D when we view ran ι_D as a subset of $L^\infty(|f^*|dm)$; so, for any such f the measures $\hat{f}d\hat{m}$ and $\hat{m}$ are equivalent.) Observing that for any measurable subset $F \subset \Upsilon$ that $\hat{m}(\bar{F}) = \hat{m}(F)$ (cf. [19, Lemma 9.4, p. 18]); one obtains that $F \subset \Upsilon$ has $\hat{m}$-measure zero if and only if there exists a nonzero $f \in H^\infty(D)$ such that F is a subset of the zero set of $\hat{f}$. So, in particular, if F is a σ-compact subset of $\Upsilon(D)$ with $\hat{m}(F)=0$, then the conclusion of Lemma 112 holds.

We now are ready to prove

Theorem 114. Let π be a unital representation of $H^\infty(D)$ into $L^\infty(\mu)$. Let ν be the minimal measure inducing π. The representation π is not one-to-one if and only if $\nu \perp \hat{m}$.

Before we justify this theorem note that Theorem 108 follows by combining the results of Theorems 114 and 106.

Proof of Theorem 114. Suppose there exists a nonzero $f \in H^\infty(D)$ such that $\pi(f)=0$. Let

$$\mu = \mu_1 + \mu_2$$

be the Lebesgue decomposition of μ with respect to m; i.e., $\mu_1 << m$ and $\mu_2 \perp m$. Clearly $\pi = \pi_1 \oplus \pi_2$ where π_i is a unital representation of $H^\infty(D)$ into $L^\infty(\mu_i)$ for i=1,2. Let ν_i be the minimal measure inducing π_i. Note $\nu_1 \perp \nu_2$, $\nu_2 \perp \hat{m}$, and $\nu = \nu_1 + \nu_2$. We need to verify $\nu_1 \perp \hat{m}$. We have reduced the problem (at hand) to the following: If π is a representation of $H^\infty(D)$ into $L^\infty(m)$ and ν is the minimal measure inducing π, then $\nu \perp \hat{m}$ provided $\ker \pi \neq 0$.

So let $\pi(f)=0$ with $f \neq 0$. Let E be a closed subset of ∂D such that $f^* |_E$ and $(\pi f)|_E$ are continuous, and $f^*|_E$ is nonvanishing. Let $z \in E$. If

$$\phi \in \mathcal{M}_z \cap \operatorname{spt}(\nu|_{E^*}),$$

then

$$\phi(f) = (\pi f)(z) = 0.$$

On the other hand, if

$$\psi \in \mathcal{M}_z \cap \operatorname{spt}(\hat{m}|_{E^*}),$$

then

$$\psi(f) = \iota_D(f)(z) = f^*(z) \neq 0.$$

Thus,

$$\operatorname{spt}(\hat{m}|_{E^*}) \cap \operatorname{spt}(\nu|_{E^*}) = \phi.$$

Now let $\{E_n\}$ be a countable collection of disjoint closed subsets of ∂D such that $(\pi f)|_{E_n}$ and $f^*|_{E_n}$ are continous, and $f^*|_{E_n}$ is nonvanishing for all n. From the last paragraph, the two sets

$$\cup \operatorname{spt}(\nu|_{E_n^*}) \text{ and } \cup \operatorname{spt}(\hat{m}|_{E_n^*})$$

are disjoint; since each is a carrier of ν and m, respectively, we see $\nu \perp \hat{m}$.

Now suppose ν is singular to $\hat{m}$. Let C be a Borel subset of $\mathcal{M}$ such that $\hat{m}(C)=0$ and $\nu(C)=1$. From the regularity of ν, there exist sequences $\{K_n\}$ and $\{L_n\}$ of compact subsets of $C\cap T$ and $C\setminus T$, respectively, such that

$$\nu(\cup K_n) = \nu(C\cap T)$$

and

$$\nu(\cup L_n) = \nu(C\setminus T).$$

Using Lemma 111 and Remark 113, we may find a (nonzero) Blaschke product B and a (nonzero) H^∞-function f such that

$$\hat{B}|_{\cup L_n} = 0$$

and

$$\hat{f}|_{\cup K_n} = 0.$$

Let g=Bf. The zero set of $\hat{g}$ is a carrier of ν, so

$$\hat{g}|_{\mathrm{spt}\ \nu} = 0.$$

From Theorem 48 it follows that $\pi(g)=0$. ∎

Obviously, we now have the following result.

Theorem 115. If G is a nicely connected region and π is a unital representation of $H^\infty(G)$ into $L^\infty(\mu)$ with ν its minimal measure, then π is one-to-one if and only if ν is not singular to $\hat{\lambda}$ if and only if π and ι_G are partially subordinate.

CHAPTER VIII

A GENERALIZATION

(OF THE RESULTS IN CHAPTER V)

In this section we present a generalization of and a different approach to the material in Chapter Five. We could have presented this chapter before Chapters 5, 6, and 7 and obtained the latter results as applications of the general theory to a special setting. We have chosen not to take this approach for two reasons; the results of this section arose from our understanding of the examples. Secondly, in exhibiting the various representations given in Chapters 5, 6, and 7 their construction depends, at least to our way of thinking, on viewing them as arising from their minimal measures rather then their associated extreme points (this latter notion to be made precise in this section).

Let A be a commutative Banach algebra with identity such that $\|a^2\|=\|a\|^2$ for each $a \in A$. Equivalently, A is a commutative Banach algebra with identity such that the Gelfand transform, $\hat{\ }$, is an isometry of A into $C(\mathcal{M}_A)$, the continuous functions on the maximal ideal space of A. We fix an element $a_0 \in A$ and let B denote the C^* algebra generated in $C(\mathcal{M}_A)$ by $\hat{a}_0$ and 1. The general theory about commutative Banach algebras (and the Gelfand transform) shows us that

$$\sigma_A(a_0)=\sigma_{C(\mathcal{M}_A)}(\hat{a}_0)=\sigma_{\mathbb{B}}(\hat{a}_0);$$

we denote this compact set by X. Using this theory again, we see the mapping from C(X) onto $\mathbb{B}$ via

$$h \to h \circ \hat{a}_0$$

is a C^* isomorphism.

<u>Problem</u> 116. Let μ be any probability measure on X. We want to characterize the continuous representations π: $A \to L^\infty(\mu)$ that send $1 \to 1$ and a_0 to χ. (Continuity is a redundant assumption; see the discussion after Question 2.)

That we choose $\pi(a_0)=\chi$ in the problem does not matter too much; we obviously have to choose the image of a_0 under π to be an element $x \in L^\infty(\mu)$ such that $\sigma_{L^\infty(\mu)}(x) \subset X$. The reader will be able to see at the end how the results of this section should be modified for the case that x is continuous and $\sigma_{C(X)}(x) \subset X$.

<u>Theorem</u> 117. Let A be any commutative Banach algebra with identity such that the Gelfand transform is an isometry. Let F be any compact space and μ a Borel measure on

F. If π is a continuous representation of A into $L^\infty(\mu)$ that sends 1 to 1, then there exists a continuous C^* representation π^* of $C(\mathcal{M}_A)$ into $L^\infty(\mu)$ such that the following diagram commutes:

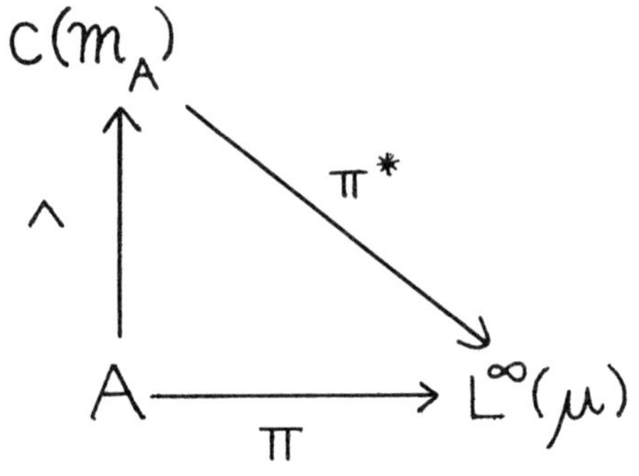

Proof. We do the obvious. The algebra

$$\{\sum_{i=1}^{n} \hat{f}_i \bar{\hat{g}}_i \colon f_i, g_i \in A\}$$

is dense in $C(\mathcal{M}_A)$ by the Stone-Weierstrauss theorem. We define π^* on this algebra by setting

$$\pi^*(\Sigma \hat{f}_i \bar{\hat{g}}_i) = \Sigma \pi(f_i)\overline{\pi(g_i)}.$$

We will next show that π^* is continuous on this subalgebra (which, in turn, shows it is well-defined and, consequently, π^* has an extension to $C(\mathcal{M}_A)$ that satisfies the conclusion of the theorem). Fix an element $\sum_{i=1}^{n} \hat{f}_i \bar{\hat{g}}_i$ of this algebra and choose $\varphi \in \mathcal{M}_{L^\infty(\mu)}$ such that

$$\varphi(\Sigma \pi(f_i)\overline{\pi(g_i)}) = \|\Sigma \pi(f_i)\overline{\pi(g_i)}\|.$$

Now we have for this φ that

$$|\varphi(\Sigma \pi(f_i)\overline{\pi(g_i)}| = |\Sigma(\varphi\circ\pi)(f_i)\overline{(\varphi\circ\pi)(g_i)}|$$

$$= |\Sigma \hat{f}_i(\varphi\circ\pi)\ \bar{\hat{g}}_i(\varphi\circ\pi)$$

$\leq \| \Sigma \hat{f}_i \bar{\hat{g}}_i \|$. ∎

Corollary 118. If A is an algebra and π is a representation satisfying the hypotheses of Theorem 116, then $\|\pi\|=1$.

Theorem 117 allows us to cast Problem 116 is a different setting; using the notation of (and before) that problem, we see the problem reduces to characterizing those representations π^* of $C(\mathcal{M}_A)$ into $L^\infty(\mu)$ such that for each $h \in C(X)$ we have

$$\pi^*(h \circ \hat{a}_0) = h.$$

That is, π^* is a representation into $L^\infty(\mu)$ that is an extension of the natural map of $\mathcal{B}$ onto C(X) (strictly speaking, we should write $C(X)|_{\mathrm{spt}\mu}$).

This last discussion suggests a more general problem (which our technique will allow us to solve).

Problem 119. Let Y, Z be compact (Hausdorff) spaces and let p be a continuous map of Y onto Z. Let μ be a regular Borel probability measure on Z. We want to characterize all the representations $\pi: C(Y) \to L^\infty(\mu)$ such that for each $h \in C(Z)$

$$\pi(h \circ p) = h.$$

Here is a special case to which the solution of the above problem applies. Let A be a commutative Banach algebra containing the constants such that $\|a^2\| = \|a\|^2$ for each $a \in A$. Let B be a subalgebra of A and let v denote the Gelfand transform of B into $C(\mathcal{M}_B)$. Let μ be a regular Borel probablity measure on $\mathcal{M}_B$. The solution to Problem 119 will allow one to understand the structure of all representations $\pi: A \to L^\infty(\mu)$ such that $\pi(1)=1$ and

$$\pi(b) = \check{b}$$

for all $b \in B$. How does 119 yield the solution to this problem? From Theorem 117 we see we need to characterize all star representations $\pi^*: C(\mathcal{M}_A) \to L^\infty(\mu)$ such that

$$\pi^*(\hat{b}) = \check{b}$$

for each $b \in B$. (A word about notation; $\hat{}$ is the Gelfand transform of A into $C(\mathcal{M}_A)$ while $\vee$ is the Gelfand transform of B into $C(\mathcal{M}_B)$.) Now we need to use the following lemma.

Lemma 120. Let Y and Z be compact (Hausdorff) spaces and let η be a multiplicative linear map from C(Z) into C(Y) with $\eta(1)=1$. Then there exists a continuous map p from Y into Z such that $\eta(f)=f \circ p$ for every $f \in C(Z)$. Consequently, η is a star representation and $\|\eta\|=1$.

Proof. Since $\sigma(\eta(h)) \subseteq \sigma(h)$ for every $h \in C(Z)$, it follows that η is continuous. If we identify Y and Z with the maximal ideal spaces of C(Y) and C(Z), then it is standard that $\eta^*(Y) \subseteq Z$, where η^* is the Banach space adjoint of η. From the definition of η^* if follows that $p \equiv \eta^* \mid Y$ satisfies the conclusion of the lemma. ∎

It is obvious that the map η in the last lemma is an isometry if and only if $p(Y)=Z$.

Returning now to the previous discussion, we see that the map $\check{b} \to \hat{b}$ for each $b \in B$ extends to a C^* isomorphism of $C(\mathcal{M}_B)$ into $C(\mathcal{M}_A)$, which we denote by η. By the last lemma there exists a continuous map p of $\mathcal{M}_A$ onto $\mathcal{M}_B$ such that $\eta(f)=f \circ p$ for every $f \in C(\mathcal{M}_B)$. Hence,

$$\hat{b}=\check{b} \circ p$$

for each $b \in B$. Its obvious now how the solution to Problem 119 yields the characterization desired.

A few remarks seem in order before we outline the solution to 119. First of all, we point out that ran p, where p is the continuous map specified in Lemma 120, has a simple characterization. Namely, if S is the closed subset of Z such that

$$\text{Ker } \eta=\{h \in C(Z): h(S)=0\},$$

then $p(Y)=S$.

Next, we would like to rephrase Problem 119 as a purely topological one. To keep our diagrams simple we make the additional assumption in Problem 119 (for this discussion only) that sptμ=Z. Obviously no harm is done in replacing $L^\infty(\mu)$ by $C(\mathcal{M}_{L^\infty(\mu)})$. Now observe that for each $\varphi \in \mathcal{M}_{L^\infty(\mu)}$ there exists a unique point $z \in Z$ such that for every $f \in C(Z)$ we have

$$\varphi(f)=f(z).$$

A moment's thought shows the map, q, just described (i.e., $\varphi \to z$) is a continuous map of $\mathcal{M}_{L^\infty(\mu)}$ onto Z that induces the natural imbedding of C(Z) into $C(\mathcal{M}_{L^\infty(\mu)})$.

Now suppose π is a representation satisfying the properties in Problem 119 (with the assumptions in the last paragraph). Let S be that closed subset of Y such that

$$\ker\pi=\{f \in C(Y): f(S)=0\}.$$

In the obvious way we identify C(Y) /Ker π and C(S) as isomorphic C^* algebras. With this identification π induces an isometry of C(S) into $C(\mathcal{M}_{L^\infty(\mu)})$. Let t be the continuous map (given by Lemma 120) of $\mathcal{M}_{L^\infty(\mu)}$ onto S that induces the latter isometry.

In addition, the hypotheses in 119 show that the map from C(Z) into C(Y) composed with the "quotient map" of C(Y) onto C(S) is an isometry. A little reflection shows this isometry is induced by the map $p|_S$. Hence, p(S)=Z.

This last discussion can now be summarized as follows: The diagram

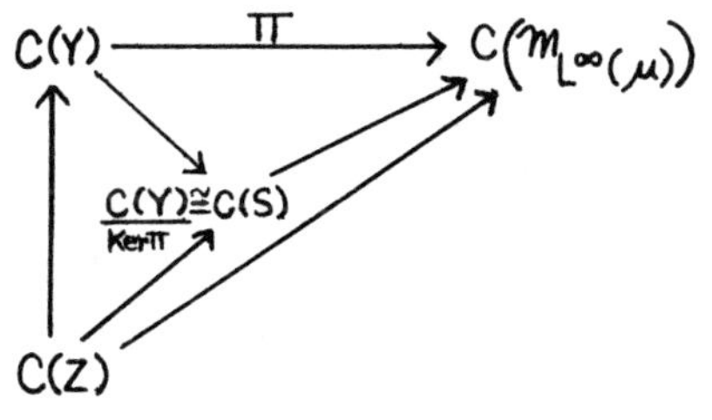

is equivalent to the diagram

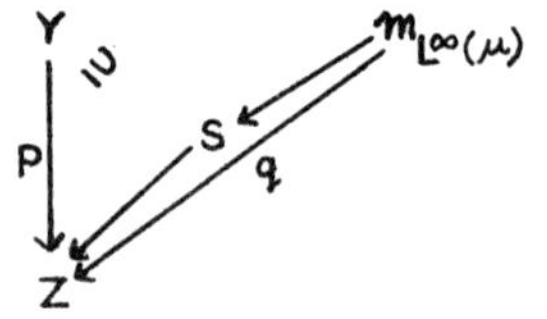

It is now clear (via Lemma 120) that Problem 119 is equivalent to finding all closed subsets $S \subset Y$ and all continuous maps d of $\mathcal{M}_{L^\infty(\mu)}$ onto S such that

$$p(S)=Z,$$

and

$$p \circ d = q.$$

(A surprising result of our solution of 119 is that for a given S that satisfies the above, there is a unique map d. However, the idea that one should seek the sets S by an appropriate minimal condition seems like the obvious thing to do.)

Lastly, before we begin, we remark that the solution to 119 we will give should be obvious to the reader who understands the results of Chapter V for the case that Z is a metrizable compact space (for then every closed subset of Z is a G_δ; a fact not necessarily true in an arbitrary compact space).

Outline of the solution to Problem 119. The compact spaces Y and Z, the continuous map p from Y onto Z, and the regular Borel probability measure μ on Z are fixed now for the rest of the paper. We let η be the C^*-isometry of C(Z) into C(Y) induced by p;

$$\eta(h) = h \circ p \text{ for all } h \in C(Z).$$

As in Chapter V, we let $R=R_\mu$ be the set of all positive Borel measures τ on Y such that

$$\int_Y \eta(h)d\tau = \int_Z h d\mu \text{ for all } h \in C(Z).$$

We note $R \neq \phi$ follows from the Hahn-Banach and Riesz representation theorems. This last equation implies that each $\tau \in R$ is a probability measure on Y.

In order to keep the analogy between the results in Chapter V and the results here, we make the following definitions: for each $z \in Z$, we set

$$Y_z = \{y \in Y: p(y) = z.\}.$$

For $E \subset Z$, let $E^* = \bigcup_{z \in E} Y_z$. If E is a measurable subset of Z, then E^* is measurable. To see this, fix $\tau \in R$ and define a measure $\widetilde{\mu}$ on Z via the equation

$$\widetilde{\mu}(E) = \tau(p^{-1}(E))$$

for all Borel sets $E \subset Z$. Consequently, if $h \in C(Z)$, then

$$\int_Z h\,d\widetilde{\mu} = \int_Y \eta(h)\,d\tau.$$

Since $\tau \in R$ we see $\widetilde{\mu} = \mu$. Hence, from the definition of $\widetilde{\mu}$ we have $\tau(E^*) = \mu(E)$ for each Borel set E. The continuity of p implies E^* is a Borel set provided E is. Now from the regularity of each of the measures, τ and μ, and the equality derived relating the two measures, we see that E^* is a set of measure zero for τ if E is a set of measure zero for μ. The desired result now follows. We have also established that $\tau \in R$ if and only if $\tau(E^*) = \mu(E)$ for each measurable set E.

From the results of the last paragraph there is a natural isometry, that we denote by η too, from $L^1(\mu)$ into $L^1(\tau)$ for any $\tau \in R$. To be sure, suppose $g \in L^1(\mu)$. We then choose a pairwise disjoint sequence $\{E_n\}$ of closed subsets of Z such that $g|_{E_n}$ is continuous and $\mu(\cup E_n) = 1$. Let $\eta(g)$ be that element in $L^1(\tau)$ defined by

$$\eta(g)(y) = g(p(y))$$

for each $y \in E_n^*$. The rest of the details are left to the reader.

It is easy to verity that R is convex and weak-star compact as a subset of $C(Y)^* = M(Y)$, the space of all regular Borel measures on Y. Therefore, from the Krein-Milman Theorem [43, Theorem 3.21] it follows that R is the closed convex hull of the set of its extreme points; i.e.,

$$R = \overline{co}(\mathrm{Ext}R).$$

Our major result describes a natural one-to-one correspondence between the set ExtR and the set of representations π satisfying the properties stated in Problem 119.

Propositon 121. Each $\tau \in R$ induces a continous linear map, L_τ, of C(Y) into $L^\infty(\mu)$ that is bounded by one and satisfies

$$L_\tau(\eta(h)f)=hL_\tau(f)$$

for each $h \in C(Z)$ and for each $f \in C(Y)$.

Proof. Let $f \in C(Y)$. Clearly $f d\tau$ is a continuous linear functonal on $L^1(\tau)$. Restricting this linear functional to the isomorphic image, $\eta(L^1(\mu))$, of $L^1(\mu)$, we see there exists a unique element $r \in L^\infty(\mu)$ such that

$$\int_Y \eta(g) f d\tau = \int_Z g r d\mu \text{ for all } g \in L^1(\mu).$$

Defining $L_\tau(f)=r$, one easily sees that L_τ is a linear map of C(Y) into $L^\infty(\mu)$. It also follows that $\|L_\tau\|=1$ because $\|r\|_\infty \leq \|f\|_\infty$ and $L_\tau(1)=1$. Now for each $g \in L^1(\mu)$ and each $h \in C(Z)$ we have

$$\int_Z g h r d\mu = \int_Y \eta(gh) f d\tau$$

$$= \int_Y \eta(g)\eta(h) f d\tau$$

$$= \int_Y \eta(g)[\eta(h)f] d\tau.$$

Hence, via the definition of L_τ, we see

$$L_\tau(\eta(h)f)=hr=hL_\tau(f). \blacksquare$$

Definition 122: The set of all linear transformations L from C(Y) into $L^\infty(\mu)$ satisfying $\|L\|=1$ and

$$L(\eta(h)f)=hL(f)$$

for every $h \in C(Z)$ and every $f \in C(Y)$, we denote by Λ. Obviously Λ is a convex set. The importance of this set is revealed in the next theorem.

Theorem 123. The mapping $\tau \in R \rightarrow L_\tau \in \Lambda$ described in Proposition 121 is a one-to-one, linear mapping of R onto Λ.

Proof. It follows immediately from the definiton of L_τ that

$$L_{\alpha\tau_1+\beta\tau_2} = \alpha L_{\tau_1} + \beta L_{\tau_2}$$

for all $\tau_1,\tau_2 \in R$ and for all nonnegative α and β satisfying $\alpha+\beta=1$. Furthermore, if $L_{\tau_1}=L_{\tau_2}$, then because $\eta(1)=1$ it follows that τ_1 and τ_2 are equal as elements of $C(Y)^*$; hence, $\tau_1=\tau_2$. What remains to be shown is that if $L\in\Lambda$, then there exists $\tau\in R$ such that $L=L_\tau$. The proof of this latter fact is similar to that of the proof of Theorem 54, with one major difference (arising from the fact Z is an arbitrary compact space).

What we shall prove here is the following: suppose $A=\{f_1,f_2,\dots,f_m\}$ is a finite subset of $C(Y)$ and E is a closed subset of Z such that $L(f_k)|_E$ is continuous for each $f_k\in A$. Then there exists a positive regular Borel measure $\tau=\tau_{E,A}$ on Y satisfying

$$\operatorname{spt}\tau \subset E^*,$$

$$\int_Y \eta(h)d\tau = \int_E h d\mu$$

for all $h\in C(Z)$, and

$$\int_Y f_k\eta(h)d\tau = \int_E L(f_k)hd\mu .$$

for all $1\le k\le m$ and all $h\in C(Z)$. With this fact it follows (by "pasting together" a countable collection of such $\tau_{E,A}$'s with $\mu(\cup E)=1$, as done in the proof of Theorem 54) that there exists a measure $\tau=\tau_A\in R$ satisfying

$$\int_Y f_k\eta(h)d\tau = \int_Z L(f_k)hd\mu$$

for all $1\le k\le m$ and all $h\in C(Z)$. The reader can then show (as done in the proof of Theorem 54) that any (weak-star) cluster point of the net $\{\tau_A\}$, say τ, has the property that $\tau\in R$ and $L=L_\tau$. We now give the details for the first aforementioned fact of this paragraph.

Let S_A be the vector space in C(Y) generated by $\eta(C(Z))$ and A. Choose a net, $\{\mu_j\}$, of positive measures each of which is supported in E,

$$\mu_j \to \mu|_E$$

weak-star as measures $(C(E)^* = M(E))$, and each μ_j is a positive linear combination of point masses on E,

$$\mu_j = \sum_{i=1}^{N_j} c_{ji}\, \delta_{z_{ji}}$$

where $z_{ji} \in E$, $c_{ji} > 0$ with $\sum_i c_{ji} = 1$, and $\delta_{z_{ji}}$ is point mass at z_{ji}. (The existence of such a choice is guaranteed by the Krein-Milman Theorem in conjunction with the characterization of the extreme points of the closed unit ball of the positive measures in M(E).) We claim that for all $t \in S_A$ and for all i and j that

(124) $|L(t)(z_{ji})| \leq \|t\|_{\{z_{ji}\}}^*$.

Suppose to the contrary the inequality in 124 is false. Without loss of generality, we may assume then that there exist some $t \in S_A$, and some i and j such that

$$L(t)(z_{ji}) = 1,$$

and

$$\|t\|_{\{z_{ji}\}}^* < 1.$$

The latter inequality allows us to choose an open set $\mathcal{O} \subset Z$ containing z_{ji} such that $\|t\|_{\mathcal{O}}^* < 1$. Now we may pick a function $q \in C(Z)$ satisfying $0 \leq q \leq 1$, $q(z_{ji}) = 1$ and $q(Z \setminus \mathcal{O}) = 0$. It follows from the properties of t and q that

$$\|\eta(q)t\| < 1.$$

Using our assumption that $L \in \Lambda$, we also have

$$L(\eta(q)t)(z_{ji}) = L(t)(z_{ji}) = 1.$$

The last inequality and the last equality imply $\|L\| > 1$, a contradiction to our assumption

$L\in\Lambda$. This establishes the inequality (124).

Now using equation (126) and the Hahn–Banach and Riesz Representation Theorems, we may choose, for each i and j, a measure $\tau_{ji}\in E^*$ satisfying

$$\tau_{ji}(\{z_{ji}\}^*)=\|\tau_{ji}\|=1,$$

and

$$L(t)(z_{ji})=\int_Y t\,d\tau_{ji}$$

for all $t\in S_A$. Let τ be (weak-star) cluster point (in $C(Y)^*$) of the net

$$\{\sum_i c_{ji}\tau_{ji}\}_j.$$

From Lemma 43 we see that spt $\tau\subset E^*$. Using the fact $\mu_j \to \mu|_E$ weak-star, we may conclude

$$\int_{E^*} \eta(h)d\tau = \int_E h\,d\mu$$

for all $h\in C(Z)$. Furthermore, for these h's and those f_k's in A, we have by using the appropriate subnet that

$$\int_Y f_k\eta(h)d\tau = \lim_j \left[\sum_i \int_{E^*} f_k\eta(g)c_{ji}d\tau_{ji}\right)$$

$$= \lim_j \int_E L(f_k)h\,d\mu_j$$

$$= \int_E L(f_k)h\,d\mu. \blacksquare$$

Corollary 125. A measure $\tau\in\mathrm{Ext}R$ if and only if $L_\tau\in\mathrm{Ext}\Lambda$.

Our attention is now turned to showing that $L_\tau\in\mathrm{Ext}\lambda$ is equivalent to the fact that L_τ is multiplicative. Once this has been done then we see that <u>the answer to problem 119 is</u>

<u>the following</u>: the representations π from $C(Y)$ into $L^\infty(\mu)$ satisfying

$$\pi(\eta(h))=h$$

for all $h \in C(Z)$ are precisely those maps L_τ where $\tau \in \text{Ext}R$.

One direction of the equivalence mentioned above is easy. If L_τ is multiplicative, then $L_\tau \in \text{Ext}\Lambda$. In fact, the next lemma shows L_τ is an extreme point of the set of all linear maps T from C(Y) in to $L^\infty(\mu)$ that have $\|T\| \leq 1$.

Lemma 126. Let Y and X be compact (Hausdorff) spaces. If T is a nonzero multiplicative linear map of C(Y) into C(X), then T is an extreme point of the set of all linear maps L from C(Y) into C(X) with $\|L\| \leq 1$.

Proof. Since $T(1)=1$ we may apply the results and the proof of Lemma 120. Let p be the continuous map of X into Y such that $T(f)=f \circ p$ for all $f \in C(Y)$. Recall from the proof of this lemma that $T^*(\delta_x)=\delta_{p(x)}$ for each $x \in X$.

Now suppose $T=tL_1+(1-t)L_2$ where $0<t<1$ and L_1 and L_2 are linear maps of C(Y) into C(X) with $\|L_1\|=\|L_2\|=1$. (Recall $\|T\|=1$.) The equality $T^*=tL_1^*+(1-t)L_2^*$, the equality $\|T\|=\|L_1\|=\|L_2\|=1$, and the fact that the extreme points of the closed unit ball of M(Y) are the point masses imply that

$$L_1^*(\delta_x)=L_2^*(\delta_x)=\delta_{p(x)}$$

for all $x \in X$. The Krein-Milman Theorem and the fact that L_1^* and L_2^* are weak-star continuous imply now that $L_1^* = L_2^* = T^*$. ∎

The converse of the last lemma is true. (One may supply the proof by modifying the proof of Theorem 127.). The answer to problem 119 is completed with the following result.

Theorem 127. If L is an extreme point of Λ, then L is multiplicative.

Proof. It follows easily from the proof of Proposition 121 (and the fact that L is induced by an element of R) that $L(f) \geq 0$ for all $f \in C(Y)$ such that $f \geq 0$. We begin now by borrowing a technique due to R.R. Phelps [44].

Fix an element $u \in C(Y)$ with $0 \leq u \leq 1$. We first show that

(128) $L(xu)=L(x)L(u)$

for all $x \in C(Y)$. To this end, define $T^{(u)}$: $C(Y) \to L^\infty(\mu)$ by setting

$$L^{(u)}(x)=L(xu)-L(x)L(u)$$

for all $x\in C(Y)$. Clearly $L^{(u)}$ is a bounded linear map and

$$L^{(u)}(\eta(h)x)=hL^{(u)}(x)$$

for all $h\in C(Z)$ and all $x\in C(Y)$. We now claim that $L+L^{(u)}$ and $L-L^{(u)}$ belong to Λ. We will verify that $T \equiv L+L^{(u)}\in\Lambda$, the other case is handled in a similar way.

Clearly,

$$T(\eta(h)x)=hT(x)$$

for all $h\in C(Z)$ and all $x\in C(Y)$. A trivial computation shows $T(1)=1$. We claim that

$$T(x)\geq 0$$

whenever $x\in C(Y)$ and $x\geq 0$. To see this, we simply compute:

$$\begin{aligned} T(x)&=L(x)+L(xu)-L(x)L(u)\\ &=L(x)(1-L(u))+L(xu)\\ &\geq 0 \end{aligned}$$

because L is a positive map (and $1-u\geq 0$). Now fix $f\in C(Y)$ and choose $\varphi\in\mathcal{M}_{L^\infty(\mu)}$ such that

$$|\varphi(T(f))| = \|TF\|.$$

Clearly $\varphi\circ T\in C(Y)^*$ and $\varphi\circ T$ is a positive linear functional. Consequently, there exists a positive regular Borel measure γ on Y such that

$$(\varphi\circ T)(t)=\int_Y t\,d\nu$$

for all $t\in C(Y)$. Since $(\varphi\circ T)(1)=1$ we see ν is a probability measure. Hence,

$$\begin{aligned} \|Tf\| &= |(\varphi\circ T)(f)|\\ &= \left|\int_Y f\,d\nu\right|\\ &\leq \|f\|. \end{aligned}$$

But f was arbitrary, hence, $\|T\|\leq 1$. It now follows $T\in\Lambda$.

From the fact $L\in\mathrm{Ext}\Lambda$ and the equality

$$L=\tfrac{1}{2}(L+L^{(u)}) + \tfrac{1}{2}(L-L^{(u)}),$$

we now deduce that $L^{(u)}=0$; that is precisely the claim in 128. It is elementary show show now that equation 128 also holds for all $u \in C(Y)$ with $u \geq 0$. Another elementary argument shows that 128 can extended to all $u \in C(V)$ such that u is real-valued. Now for any two elements x and u in C(Y) we have

$$L(xu)=L(x\mathrm{Re}u+i\mathrm{Im}u))$$

$$=L(x)L(\mathrm{Re}u)+iL(x)L(\mathrm{Im}u)$$

$$=L(x)L(u). \quad \blacksquare$$

An answer to problem 119 has been given. One might claim (in point of fact, a person did after one of the authors presented this material in a seminar) the answer has just restated the problem in another language. We do not feel it is that inadequate. In fact, the theory in this chapter is adequate enough to answer the question asked in [8] that motivated our original inquiry into these matters. We leave it as an exercise to the reader (using only the results of this chapter up till now) there are infinitely many unital representations of $H^{\infty}(D)$ into $L^{\infty}(m)$ that send χ to χ. However, it is not apparent that this chapter shows how to construct the representations given in examples 40, 68, and 77.

A little more work though will overcome this difficulty. One may introduce a partial ordering $\leq$ on $\mathcal{R}$, the equivalence classes of R under $\leq$, identical to that given in Chapter V. Maximal elements exist in $\mathcal{R}$ and each of their equivalence classes consist of a singleton measure $\nu \in R$. These minimal measures ν are precisely the extreme points of R. (One way to justify this is that the analogue of Theorem 58 to this setting is valid.) When this is done (we leave the details to the interested reader) the "geometry" of ExtR is more apparent and the construction of examples (like those mentioned in the last paragraph) can be done. (The fact that Z is an arbitrary compact space rather than a separable compact space requires the arguments to be a little more measure theoretic.)

REFERENCES

1. P. Ahern, On the generalized F. and M. Riesz theorem, Pacific J. Math. 15 (1965), 373–376.

2. C. Apostol and B. Chevreau, On M–spectral sets and rationally invariant subspaces, J. Operator Theory 7 (1982), 247–266.

3. S. Axler, Multiplication operators on Bergman spaces, J. Reine Angew. Math. 226 (1982), 26–44.

4. S. Axler, J. Conway and G. McDonald, Toeplitz operators on Bergman spaces, (to appear) Canadian J. Math.

5. S. Brown, Some invariant subspaces for subnormal operators, Integral Eqs. Op. Theory 1 (1978), 310–333.

6. S. Brown, B. Chevreau, C. Pearcy, Contractions with rich spectrum have invariant subspaces, J. Operator Theory 1 (1979), 123–136.

7. J. Chaumat, Adherence faible etoile d'algebres de fractions rationelles, Ann. Inst. Fourier 24 (1974), 93–120.

8. B. Chevreau, C. Pearcy and A. Shields, Finitely connected domains, representations of $H^\infty(G)$, and invariant subspaces, J. Operator Theory 6 (1981), 375–405.

9. B. Cole and T. Gamelin, Tight uniform algebras and algebras of analytic functions, J. Functional Anal. 46 (1982), 158–220.

10. J. Conway, Subnormal Operators, Pitman Publishing Inc., Research notes in math. 51, Marshfield, Mass., 1981.

11. J. Conway and R. Olin, A functional calculus for subnormal operators, Memoirs Amer. Math. Soc. 184 (1977), vii+61.

12. H. Dales, Automatic continuity; A survey, Bull. London Math. Soc. 10 (1978), 129–183.

13. A. Davie, Dirichlet algebras of analytic functions, J. Junctional Anal. 6 (1970), 348–356.

14. A. Davie, Bounded limits of analytic functions, Proc. Amer. Math. Soc. 32 (1972) 127–133.

15. J. Dudziak, Spectral mapping theorems for subnormal operators, Ph.D thesis (1981), Indiana University.

16. P. Duren, Theory of H^p–spaces, Academic Press, New York, 1970.

17. P. Fillmore, J. Stampfli and J. Williams, On th essential numerical range, the essential spectral and problem of Halmos, Acta Sci. Math. (Szeged) 33(1972), 179–191.

18. S. Fisher, Function Theory on Planar Domains, Wiley–Interscience Pub., New York, 1983.

19. T. Gamelin, Uniform Algebras, Prentice-Hall, Englewood Cliffs, N.J. 1969.

20. T. Gamelin, Lectures on $H^\infty(D)$, Notas de Matematika No 21, Univ. Nacion, de la Plata, 1972.

21. T. Gamelin, Rational Approximation Theory, Course Notes, Univ. of Calif. at Los Angeles, Fall 1975.

22. T. Gamelin and J. Garnett, Distinguished homomorphisms and fiber algebras, Amer. J. Math. SCLL (1970), 455-474.

23. T. Gamelin and J. Garnett, Pointwise bounded approximation and hypodirichlet algebras, Bull. Amer. Math. Soc. 77 (1971), 137-141.

24. T. Gamelin and J. Garnett, Pointwise bounded approximation and Dirichlet algebras, J. Functional Anal. 8 (1971), 360-404.

25. J. Garnett, Bounded Analytic Functions, Academic Press, New York, 1981.

26. I. Glicksberg, Dominating representing measures and rational approximation, Trans. Amer. Math. Soc. 130(1968), 425-462.

27. P. Halmos, G. Lumer and J. Schaffer, Square roots of operators, Proc. Amer. Math. Soc. 4(1953), 143-149.

28. K. Hoffman, Banach Spaces of Analytic Functions, Prentice-Hall, Englewood Cliffs, N.J., 1962.

29. D. Newman, Some remarks on the maximal ideal space structure of H^∞, Annals of Math. 70(1959), 438-445.

30. B. Sz-Nagy and D. Fois, Harmonic Analysis of Operator son Hilbert Space, North-Holland, Amsterdam, 1970.

31. R. Olin, Functional relationships between a subnormal operator and its minimal normal extension, Pac. J. Math. 63(1976), 221-229.

32. R. Olin and J. Thomson, Lifting the commutant of a subnormal operator, Canadian J. Math. 31(1979), 148-156.

33. R. Olin and J. Thomson, Algebras of subnormal operators, J. Functional Anal. 37(1980), 271-301.

34. C. Pearcy, Some recent developments in operator theory, Conference board of the mathematical sciences no. 36., Amer. Math. Soc. (1975), v+73.

35. H. Radjavi and P. Rosenthal, Invariant Subspaces, Springer-Verlag, New York, 1973.

36. L. Rubel and A. Shields, The space of bounded analytic functions on a region, Ann. Inst. Fourier (Grenoble) 1 (1966), 235-277.

37. D. Sarason, A remark on the weak-star topology of ℓ^∞, Studia Math. 30 (1968), 355-359.

38. D. Sarason, Weak-star density of polynomials, J. Reine Angew. Math. 252 (1972), 1-15.

39. A. G. Vitushkin, Analytic capacity of sets in problems of approximation theory, Russian Math Surveys 22 (1968), 139–200.

40. A. Wilansky, Functional Analysis, Blaisdell Pub. Co., New York, 1964.

41. T. Yoshino, Subnormal operators with a cyclic vector, Tohoku Math. 21 (1969), 47–55.

42. W. Arveson, An Invitation to C^*–Algebra, Springer–Verlag, New York, 1976.

43. W. Rudin, Functional Analysis, McGraw–Hill, New York, 1973.

44. R. R. Phelps, Extremal operators and homomorphisms, Trans. Amer. Math. Soc. 108 (1963), 265–274.

R. Olin and J. Thomson
Dept. of Math
Virginia Tech
Blacksburg, VA 24061

T. L. Miller
Dept of Math
Drawer MA
Mississippi State U.
Mississippi State, MS 39762

General instructions to authors for
PREPARING REPRODUCTION COPY FOR MEMOIRS

For more detailed instructions send for AMS booklet, "A Guide for Authors of Memoirs." Write to Editorial Offices, American Mathematical Society, P. O. Box 6248, Providence, R. I. 02940.

MEMOIRS are printed by photo-offset from camera copy fully prepared by the author. This means that, except for a reduction in size of 20 to 30%, the finished book will look exactly like the copy submitted. Thus the author will want to use a good quality typewriter with a new, medium-inked black ribbon, and submit clean copy on the appropriate model paper.

Model Paper, provided at no cost by the AMS, is paper marked with blue lines that confine the copy to the appropriate size. Author should specify, when ordering, whether typewriter to be used has PICA-size (10 characters to the inch) or ELITE-size type (12 characters to the inch).

Line Spacing – For best appearance, and economy, a typewriter equipped with a half-space ratchet – 12 notches to the inch – should be used. (This may be purchased and attached at small cost.) Three notches make the desired spacing, which is equivalent to 1-1/2 ordinary single spaces. Where copy has a great many subscripts and superscripts, however, double spacing should be used.

Special Characters may be filled in carefully freehand, using dense black ink, or INSTANT ("rub-on") LETTERING may be used. AMS has a sheet of several hundred most-used symbols and letters which may be purchased for $5.

Diagrams may be drawn in black ink either directly on the model sheet, or on a separate sheet and pasted with rubber cement into spaces left for them in the text. Ballpoint pen is *not* acceptable.

Page Headings (Running Heads) should be centered, in CAPITAL LETTERS (preferably), at the top of the page – just above the blue line and touching it.

LEFT-hand, EVEN-numbered pages should be headed with the AUTHOR'S NAME;

RIGHT-hand, ODD-numbered pages should be headed with the TITLE of the paper (in shortened form if necessary).

Exceptions: PAGE 1 and any other page that carries a display title require NO RUNNING HEADS.

Page Numbers should be at the top of the page, on the same line with the running heads.

LEFT-hand, EVEN numbers – flush with left margin;

RIGHT-hand, ODD numbers – flush with right margin.

Exceptions: PAGE 1 and any other page that carries a display title should have page number, centered below the text, on blue line provided.

FRONT MATTER PAGES should be numbered with Roman numerals (lower case), positioned below text in same manner as described above.

MEMOIRS FORMAT

It is suggested that the material be arranged in pages as indicated below.
Note: <u>Starred items</u> (*) <u>are requirements of publication</u>.

Front Matter (first pages in book, preceding main body of text).

Page i – *Title, *Author's name.

Page iii – Table of contents.

Page iv – *Abstract (at least 1 sentence and at most 300 words).

*<u>*1980 Mathematics Subject Classification (1985 Revision)</u>*. This classification represents the primary and secondary subjects of the paper, and the scheme can be found in Annual Subject Indexes of MATHEMATICAL REVIEWS beginning in 1984.

Key words and phrases, if desired. (A list which covers the content of the paper adequately enough to be useful for an information retrieval system.)

Page v, etc. – Preface, introduction, or any other matter not belonging in body of text.

Page 1 – Chapter Title (dropped 1 inch from top line, and centered).

Beginning of Text.

Footnotes: *Received by the editor date.

Support information – grants, credits, etc.

Last Page (at bottom) – Author's affiliation.

ABCDEFGHIJ – 89876